Pelican Books

Metals in the Service of Man

Dr Arthur Street and Professor William Alexander graduated in the early 1930s from Birmingham University, where they studied metallurgy and conducted researches. Until his retirement in 1975, Dr Street was Chairman of a well-known diecasting company. He wrote one of the first books about diecasting and has recently produced two modern ones. In 1975 he became the first European to win the Doehler Award, given annually by the American Die Casting Institute. His principal hobbies are grand opera, foreign travel and lecturing on art, metals and music.

For many years Professor Alexander was connected with production and metallurgical research in one of the largest non-ferrous metal organizations in Britain, where at various times he worked on copper, aluminium, lead, zinc, iron, uranium, thorium, niobium, beryllium and some of their alloys. After the Battle of Britain he solved in six weeks the problem of cracking in submarine periscope tubes which had been baffling metallurgists for many years. From 1967 until his retirement in 1976 he was Professor of Metallurgy at the University of Aston in Birmingham, where he concentrated on developing course work on industrial metallurgy, heat treatment and engineering materials. Now Professor Emeritus, he continues metallurgical consulting and research work on safety problems with molten metals. In 1982 he became the first recipient of the Institution of Metallurgists' Professional Service Award. Professor Alexander is President of the famous athletic club, The Birchfield Harriers.

METALS IN THE SERVICE OF MAN

Ninth Edition

Arthur Street

William Alexander

Dear Philip

I was delighted that you got a copy of this book and enjoyed reading it. You will gather from the list of editions that we have to revise the book every so often — I have just sent Penguin another revision. So the activities with 'metals' keeps me in touch with modern developments

Yours
Arthur

[I hope you can decypher the writing!!]

PENGUIN BOOKS

PENGUIN BOOKS

Published by the Penguin Group
27 Wrights Lane, London W8 5TZ, England
Viking Penguin Inc., 40 West 23rd Street, New York, New York 10010, USA
Penguin Books Australia Ltd, Ringwood, Victoria, Australia
Penguin Books Canada Ltd, 2801 John Street, Markham, Ontario, Canada L3R 1B4
Penguin Books (NZ) Ltd, 182–190 Wairau Road, Auckland 10, New Zealand

Penguin Books Ltd, Registered Offices: Harmondsworth, Middlesex, England

First published 1944
Second edition 1951
Third edition 1954
Fourth edition 1962
Reprinted with revisions 1964
Reprinted with revisions 1968
10 9 8 7 6 5 4 3 2 1

Fifth edition 1972
Sixth edition 1976
Seventh edition 1979
Eighth edition 1982
Reprinted with revisions 1985
Ninth edition 1989

Made and printed in Great Britain
by Richard Clay Ltd, Bungay, Suffolk
Filmset in Monophoto Plantin

To Brenda and Kath, with many thanks for their help, encouragement and endurance since the first edition of this book was written

CONTENTS

PREFACE

The science of metals is a specialized one and tends to be shut away from the general reader in rather grim-looking text books and papers. This is a pity, because metals retain the attraction and wonder they had for us in our youth and they are of great importance to mankind. We hope this book will interest all those who handle metals in their leisure, pleasure, or daily work, whether this is in the home, industry, school or university. We have tried to avoid the 'sensational discovery' style and have not hesitated to include discussions of some complex subjects, while still attempting to present our material in a readable form. Specialized technical terms have been defined in the glossary.

Whenever this book has been revised or re-written it has been necessary to decide which of the many developments in metallurgy warrant inclusion. Ideas for new processes have at first been thought to be fantastic, then impossible, then diagnosed as possible but impracticable. Finally the processes have been proved to be practicable and then have entered into the repertoire of the production metallurgist. We have always been impressed and often astonished at the ingenuity, skill and perseverance of those who have made new discoveries or developed new processes, thereby enlarging the scope of metals in the service of mankind.

Thanks to 'a little help from our friends' it has been possible to bring the book up to date for each new edition and reprint. We are indebted to members of the Institute of Metals, various development associations, research institutions, universities and the staffs of metallurgical companies in many parts of the world. Readers have kindly written to us with suggestions for improvements – and sometimes with corrections. We thank all these and remain sincerely grateful to the friends who, in the early nineteen-forties, encouraged us to embark on the first edition of this book which, we understand, has persuaded many young people that metallurgy as a career is interesting and rewarding.

A.C.S., W.O.A

'DRAMATIS PERSONAE'

Principals

IRON	The most important metal
ALUMINIUM	The light metal, second in importance to iron
COPPER	The conductivity metal
ZINC	The galvanizing metal
LEAD	The battery metal
TIN	The metal that tins the can
NICKEL	The versatile metal

Supporting characters

MAGNESIUM, BERYLLIUM	The ultra-light duet
TITANIUM	The strong middleweight
CHROMIUM	The stainless metal
TUNGSTEN	The dart players' 'light' heavyweight
MANGANESE, VANADIUM	The scavenging metals
BORON, COBALT, MOLYBDENUM	A non-metal and two metals that give improved properties to steel
NIOBIUM (alias COLUMBIUM), ZIRCONIUM	Two important new arrivals
CADMIUM	The weather-resister
TANTALUM	The capacitor metal
GOLD, SILVER, PLATINUM	The precious trio
IRIDIUM, PALLADIUM, RHODIUM, RUTHENIUM	The valuable quartet
BARIUM, CAESIUM, CALCIUM, POTASSIUM, SODIUM, STRONTIUM	The reactive sextet

OSMIUM	The heaviest metal
LITHIUM	The lightest metal
MERCURY	The liquid metal
GALLIUM, HAFNIUM, INDIUM, RHENIUM, RUBIDIUM, SCANDIUM, THALLUM, YTTRIUM	Some rare metals
CERIUM, DYSPROSIUM, ERBIUM, EUROPIUM, GADOLINIUM, HOLMIUM, LANTHANUM, LUTETIUM, NEODYMIUM, PRASEODYMIUM, PROMETHIUM, SAMARIUM, TERBIUM, THULIUM, YTTERBIUM	The fifteen 'rare earth' metals, known as Lanthanides
PLUTONIUM, RADIUM, THORIUM, URANIUM	Radioactive metals
ACTINIUM, AMERICIUM, BERKELIUM, CALIFORNIUM, CURIUM, EINSTEINIUM, FERMIUM, LAWRENCIUM, MENDELEVIUM, NEPTUNIUM, NOBELIUM, PROTOACTINIUM (The derivation of most of these characters' names will be obvious)	Very rare metals, most of them synthetic products of radioactivity. They are classed as Actinides, because the one with the lowest atomic number is actinium. This group also contains uranium, thorium and plutonium, which play a more active part than the other Actinides
FRANCIUM, POLONIUM, TECHNETIUM	Unstable metals, formed in the disintegration of radioactive elements
ANTIMONY, ARSENIC, BISMUTH, TELLURIUM	Semi-metals or metalloids
GERMANIUM, SELENIUM, SILICON	Semi-conductor elements
NITROGEN	A gas that helps to harden some metals
CARBON, OXYGEN	Two non-metals, vital to life and useful in metallurgy

LIST OF PLATES

A NOTE ON METRICATION

In 1965 the British Government acceded to industry's request for the U.K. to adopt metric units and the decision was also taken to adopt a particular version of the metric system, the International System of Units (S I). In 1971 the currency of Britain went decimal; engineering, commerce and education were already 'thinking metric' and using the international S I units. We therefore changed the measurements in this book to metric and we began to use the S I units, with sufficient information about the conversions to help those who still found it difficult to visualize the strength of metals in terms of newtons per square millimetre. In conversion from one system of measurement to another some approximations are necessary; for example one pound equals 0·45 359 237 kg but an approximate figure is given below.

1 pound	equals	0·454 kilogram
1 inch	equals	25·4 millimetres
1 gallon	equals	4·5 litres
1 ton force	equals	15·44 newtons per sq. millimetre
per sq. inch		(usually expressed as 15·44 N/mm²)

The derivation of newtons, in connection with the tensile strength of metals, is discussed on page 93. We have also shown the strength of metals in tons per square inch for comparison, in several parts of the book.

TONS OR TONNES?

The 1000 kilogram metric tonne is equal to 2204 pounds and is therefore about 1·6 per cent lighter than an Imperial ton. Most industrial concerns are already purchasing metal by the tonne, although 'the man in the street' may remain reluctant to use the word, owing to its 'olde worlde' flavour. We have therefore compromised. Where a precise weight is indicated, for example in referring to the cost of metals or modern production statistics, we have used 'tonnes'. Where we mention the annual production of metal before metrication or give a fact about the early use of metals, such as the weight of wrought iron in the Eiffel Tower, we have retained 'tons'.

DISTANCES

For long distances, no great precision was needed, so we used the motorist's conversion of five miles to eight kilometres. For precise measurements we convert inches to millimetres with a divisor of 25·4, or to centimetres with a divisor of 2·54. Intermediate measurements have been converted to metres.

For very small measurements, the micron (a thousandth of a millimetre) or the Ångström unit (a ten-millionth of a millimetre) have been used but now scientists prefer the nanometre (one millionth of a millimetre). Thus 10 Ångström units equal one nanometre.

TEMPERATURES

Temperatures in European industrial processes are generally expressed in degrees Celsius, after the Swedish scientist Anders Celsius. In Britain these temperature units are sometimes called Centigrade. We are still learning to use the Celsius–Centigrade scale in discussing the weather and cookery; in the U.S.A. Fahrenheit is used quite widely in industry. The scientific unit of temperature is the Kelvin (K), whose zero is absolute zero but whose intervals are the same as those of Celsius–Centigrade. A temperature expressed in degrees C is equal to the temperature in degrees K minus 273·15 degrees. Thus the boiling point of water is 373·15 K.

PRESSURES

The S I unit for pressure is the Pascal (Pa), which equals one newton per sq. metre (N/m^2). One ton per sq. in. converts to 15·44 megapascals (MPa), a megapascal being one million pascals, or 15·44 newtons per sq. mm (N/mm^2).

GENERAL

For approximate measurements we have converted approximately. Thus in early editions some Roman nails were described as 16 ins. long but they become 400 mm long, after metrication. As most plant engineers still use horse power we retained that. In referring to telescope mirrors we have used inches for an American instrument because we gather it is still referred to in that way, but British telescopes have now 'gone metric'. So far as we are aware the hour will be divided into sixty minutes and the minute into sixty seconds for a long time to come and the local pub will be serving pints for several more years.

1

METALS AND CIVILIZATION

Although metal-working was not the first craft known to mankind, our present material civilization is derived from the knowledge and use of metals. From the time of Tubal-cain* onwards, the early users of metals would probably have described them as bright substances which were hard but could be hammered into various useful shapes. All the metals known to them were heavy – iron, for instance, being nearly three times as heavy as granite. If the metal-worker were also a hunter or warrior he might have spoken with satisfaction of the sharpness of metal weapons and of their endurance through many combats. Craftsmen in wood or stone would have praised the new metal cutting-tools which made their jobs so much easier.

Man learned to make fires hot enough to melt metals in earthenware containers, which the Romans called 'crucibuli' and which we now call 'crucibles'. Molten metal was poured into the cavity made by placing together two halves of a hollowed-out clay or stone mould: the metal filled the cavity and, when solid, it was found to have taken the shape of the cavity. Archaeologists have discovered ancient bronze swords and arrowheads made by casting in this manner.

The art of blending metals was gradually developed and some alloys formed in this way were stronger, harder, and tougher than the metals of which they were composed. Probably the first alloy to be made was a bronze, consisting of copper with about one part in ten of tin. The primitive metallurgists discovered that if a greater proportion of tin was used the alloy was harder, while less tin gave a softer alloy. So for different purposes bronzes with varying tin contents were deliberately produced. By the time the Romans came to Britain, they were using iron and bronze for weapons, tools, and farming implements; copper for

* Genesis iv, 22.

vessels and ornaments; lead for water pipes, baths, and even coffins; tin, gold, and silver for ornaments; and silver, brass, and bronze for coinage.

Gold and silver were called 'noble metals' because they could be exposed to the atmosphere for a long time without tarnishing and because they could be melted repeatedly without much loss in weight. These characteristics led to their being used for jewellery and coinage. The possession of noble metals consequently became a measure of wealth so that gold and silver were coveted for their monetary as distinct from their decorative value. All the other metals then known, such as tin, lead, copper, and iron, were considered 'base metals'.

Owing to the general similarities between all metals it was perhaps natural to imagine that one metal could be changed into another, and it seemed particularly desirable that base metals should be transmuted into noble ones. The mystic pseudo-science of alchemy emerged from China and Egypt in the third century B.C.; it spread to Europe and thrived there until the eighteenth century. Alchemy became associated with the fascinating possibility of converting base metals into gold or silver, through the agency of a secret substance called the elixir, the grand magistry or the philosophers' stone.

As recently as the end of the eighteenth century a writer defined alchemy as 'a science and art of making a fermentative powder which transmutes imperfect metals into gold and which serves as a universal remedy for all the ills of man, animals, and plants'. The philosopher Francis Bacon was nearer the truth when he commented: 'Alchemy may be compared to the man who told his sons that he had left them gold buried somewhere in his vineyard; where they by digging found no gold, but by turning up the mould about the roots of the vines, procured a plentiful vintage. So the search and endeavours to make gold have brought many useful inventions and instructive experiments to light.'

While the alchemists and their wealthy, and often exasperated, patrons were trying all kinds of experiments to produce the philosophers' stone, the metal-workers, though using fewer incantations, were developing almost as wonderful a process – the changing of dull, earthy minerals or 'ores' into metals by smelting them with charcoal in a fire or furnace. They learned how to recognize those metallic ores which could be smelted profitably and how to transform them into metal. It is not surprising that their efforts were sometimes unsuccessful. Even today, producers of metal encounter difficulties because of impurities in the ore that upset smelting operations or have a harmful influence on the resulting metal. In those early days such happenings were freely attributed to the Evil Eye, or to the attention of hobgoblins, or Old Nick himself. The name of nickel was derived from this, while the name for cobalt originated from the German elves called *kobolds*.

Between the Middle Ages and the beginning of the industrial era, the main progress in the art of making metals was the building of larger and more efficient furnaces to produce metal in greater quantity. In 1740, when Dr Johnson used to drink tea with Sir Joshua Reynolds and Mr Garrick, Britain was the world's greatest producer of metals. We were at war with Spain and made about twenty thousand tons of iron per year.* (Now we make much more than that amount each day.) A hundred years later, when the penny post had just been introduced, Britain made one and a quarter million tons of iron annually, and by the end of the nineteenth century the figure had risen to about nine million tons per year.

Twenty years after the beginning of Victoria's reign the manufacture of steel was becoming a major industry; in 1856 Henry Bessemer made public his process for converting large quantities of pig iron into steel. The rapid development of the use of iron and steel throughout the civilized world was outstanding, though copper, tin, lead and zinc were also being produced in increasing quantities. Metals in general, and steel in particular, came to be used for making bridges, railway lines, ships, guns, implements of all kinds, and, towards the end of the nineteenth century, those noisy 'horse-less carriages' which chugged their way along the roads at a rattling speed sometimes exceeding five miles per hour.

Queen Victoria's reign extended over a period during which the *art* of metal-working was growing into the *science* of metallurgy. In 1861 Professor Henry C. Sorby of Sheffield initiated the microscopic examination of metals and thus laid the foundations of metallography. Properties of metals – their melting points, strength, hardness and electrical conductivity – were studied and correlated.

The nineteenth century must have been an exciting time for chemists and metallurgists as they gradually came to realize that many new metallic elements were waiting to be discovered. One compound would be separated from the others in a mineral on the suspicion that it contained a previously unrecognized element. The properties of the compound would be studied and a guess made at the element's atomic weight. Then many processes would be tried in the hope of isolating the metal; sometimes the experiments continued over many years. At last the metal would be revealed but often its properties would not appear impressive because it was impure and contaminated. In metals ranging

---

* The metallurgical beginnings of the Industrial Revolution are splendidly displayed at the open-air museum, near the famous iron bridge, in Telford, between Birmingham and Shrewsbury.

Table 1. Average annual world production of the common metals
      [Millions of tonnes]

	1858–1907	1908–57	1958–70	1971–81	1987
Copper	0·26	1·70	6·00	8·25	8·40
Lead	0·50	1·40	2·50	3·07	3·38
Tin	0·06	0·14	0·15	0·18	0·18
Zinc	0·30	1·40	3·30	5·82	6·80
Aluminium	0·002	0·76	8·00	12·00	15·90
Magnesium	NIL	0·04	0·15	0·25	0·38
Nickel	0·004	0·03	0·46	0·60	0·77
Steel	11·00	120·00	475·00	680·00	735.00

from nickel to vanadium the first results obtained were brittle. It needed someone with enough faith and determination to try to improve the purity and properties of the new metal. Next the scientists would wonder how the new element might be used commercially. What application could be found for a new metal which appeared to be much lighter than any of the other metals in common use? We of the twentieth century are so accustomed to seeing aluminium in all walks of life that we find it hard to imagine the feelings of the first men to produce some rather dirty-looking specks of aluminium in the early decades of the nineteenth century.

In the twentieth century the expansion of the use of metals has been so rapid that considerably more metal has been extracted during the last eighty years than during the ages from the beginning of man's history till A.D. 1900. *Table 1* shows a comparison of the world average annual production of some well-known metals, first for two fifty-year periods – 1858–1907 and 1908–57, then during the periods 1958–70 and 1971–81 and finally for one recent year, 1987.

How many people realize that our material civilization and every amenity of life depend on the work of the metallurgist and on his ability to produce the right metal for each particular purpose? In the morning we press an electric switch and current passes along a copper wire to light a lamp with a tungsten filament. We wash in water that has come through copper pipes and runs into the basin from a nickel and chromium plated brass tap. We shave with a stainless steel razor-blade, whilst anticipating breakfast that is to be cooked on a pressed steel cooker; meanwhile tea is being made with water heated in an aluminium kettle. We eat our breakfast with the aid of a stainless steel knife and fork and then drive to work in a car which is 80 per cent metal, or catch a train or bus, paying with a 50p coin made of an alloy of copper and nickel, plus some smaller coins of copper–zinc–tin alloy. This is only the beginning of a day during which hundreds of metal objects may be used, whether we operate a lathe or a typewriter, drive a tractor or observe the universe through a telescope.

Our material civilization depends on the efficient harnessing of power, but its control is made possible by the use of metals and alloys. Without metals no aircraft, automobile, military tank, electric motor or interplanetary space vehicle could operate. On a smaller scale it is difficult to imagine kitchens, typewriters, sewing kits or light fittings without metals. Some of man's latest and greatest technical achievements are structures made of at least 95 per cent metal, such as the space shuttles, super- and subsonic aircraft ranging from Concorde and the Tornado to the Airbus, the great single-span suspension bridge over the Humber and the new one over the Bosphorus.

The variety of metals which is now available has certainly benefited mankind, but it has made discrimination necessary to get the best use from each metal or alloy. For example, Concorde was designed to withstand the arduous conditions of aerodynamic heating, coupled with the long life of 45 000 flying hours required of the first supersonic aircraft for airline use. Seventy-one per cent of the structure is made of a special aluminium alloy, RR.58, which can withstand these conditions. Sixteen per cent of the weight of Concorde is of high strength steel, for the undercarriage. Titanium alloys, amounting to 4 per cent, are used in the engine nacelle; this has to withstand a temperature of about 400°C; such a temperature is too high for aluminium alloys, but not so high that heat-resisting steels, or nickel alloys, would have been required. The remaining 9 per cent comprises nickel alloys, plastics, glass and other materials.

During the last thirty years many metals have emerged from being laboratory curiosities to a position of some importance: tantalum, titanium, niobium, molybdenum, zirconium; these and many others have extended the repertoire of the metallurgist. The requirements of nuclear energy, jet aircraft and space ships have provided incentives to overcome the immense problems of winning these metals from their ores and of fabricating them. Sometimes completely new methods have been developed to shape metals that, a generation ago, were thought quite intractable.

A rigid classification of the elements into metals, such as iron and copper, and non-metals, such as carbon and boron, is not completely valid. Elements such as antimony have properties characteristic of non-metals, with some metallic tendencies; they are known as semi-metals or metalloids. Another group, known as semi-conductors, includes silicon and germanium; they have a low electrical conductivity characteristic of non-metals but the conductivity increases with increase of temperature; furthermore their conductivity can be controlled by the addition of minute quantities of impurities, a phenomenon that led to the development of transistors.

While exploitation of the new metals is proceeding, fresh endeavours are being made to widen the scope of the well-known materials. New ways of producing steel, new alloys, greater purity of metal to reveal unexpected properties, improved methods of casting, joining and cutting metals, are just a few of the developments which have taken place in the last quarter century. In these and other ways the designer and engineer are being offered a wide range of metallic materials. Using the specialized knowledge and advice of the metallurgist they are able to select the most suitable metal for the job and the most suitable treatment for that metal, bearing in mind its mechanical and physical properties, its ease of fabrication, availability and cost.

Modern civilized mankind is so familiar with metals that the difficulty of defining a metal may be overlooked. Indeed, different branches of science have different definitions. A chemist might say 'Metallic elements are those which possess alkaline hydroxides.' A physicist might define a metal as 'An element with a high electrical conductivity which decreases slightly with increase of temperature.' Someone who works with metals would remind us that they possess lustre when polished and are good conductors of electricity and heat; he would add that most of them are denser, stronger, and more malleable than the other, non-metallic, elements. Engineers will tell you that metals and alloys have great strength and capability of withstanding limited overloading without catastrophic failure, in other words, they are tough.

We, the authors, having handled a variety of metals and alloys for over fifty years, can tell you of our abiding life interest. We can reassure those following that there is still much scope for craft and ingenuity in metallurgy, and that real fundamental understanding of many of the properties of metals still calls for intensive scientific work.

2

HOW WE GET OUR METALS

When Commander Robert E. Peary was exploring Greenland in 1894 an Eskimo took him to a place near Cape York where he found three metallic meteorite masses. The Eskimos called them Saviksue, or 'the great irons', each having a name suggested by its shape. Peary removed 'The Woman' and 'The Dog' to his ship in 1895, and brought them to the American Museum of Natural History. The largest meteorite, 'The Tent', was much more difficult to remove, but in 1897 Peary and his men transferred the thirty-four ton mass to his ship. His four-year-old daughter, Marie Ahnighito Peary, threw a bottle of wine on the meteorite, which was then re-named Ahnighito. In 1904 it was hauled slowly along Broadway and 77th Street to the entrance of the American Museum. When the Hayden Planetarium was opened in 1935, Ahnighito made its last journey, halfway around Manhattan Square to the Planetarium first floor, where it has remained on show ever since.

Meteorites usually consist of an alloy of iron with about 8 per cent of nickel, with a small amount of cobalt. No doubt primitive man, whose local culture was thus by accident raised from the level of the stone age to that of the iron age, thought metallic meteorites were valuable gifts from the gods. Nowadays, however, meteorites are hardly regarded as a useful source of iron. For one thing the delivery service is erratic and the unheralded arrival of a meteorite in one's back garden would be more embarrassing than profitable.

Copper has also been discovered occasionally in the metallic form. The largest mass of pure copper was found in 1856 in what was called the Minnesota Mine, in the upper Michigan peninsula. It weighed about 500 tons and was so large that it had to be cut into pieces before being hoisted to the surface. Such a find is, however, of little significance as a source of copper, of which over eight million tonnes is made annually. Most metals, with the exception of some precious ones, usually appear in nature as minerals or ores, where they exist in chemical combination with other elements. These ores do not resemble the metals which can be extracted from them; one would hardly imagine that strong and bright metals could be produced from such uninteresting-looking earthy substances. Yet it is easy to realize that rust is a chemical compound of iron, while verdigris is a compound of copper; these are the kinds of metal-bearing materials which mineral deposits contain. The ores are treated by fire, chemical or electrical processes, to convert them into metals, these operations being known as 'smelting'. Before that can be done, the ore deposits must be mined or quarried, concentrated and transported to the smelting works.

METALLIC ORES

It is believed that the central core of the earth, about 6500 km in diameter, consists of a molten alloy of iron and nickel, at a temperature of 3700°C. Perhaps one day this may provide a source of molten metal, like a Texan oil gusher, but at the moment there would be problems. Mining metallurgists win their ores from the outer crust of the earth.

In 1924 two American scientists, F. W. Clarke and H. S. Washington, attempted to estimate the proportions of the chemical elements contained in the outer 16 km of the earth's crust. To find the approximate composition of the 17 000 000 000 000 000 000 tons comprising this outer crust, a series of averages was taken, based on 5159 chemical analyses of minerals all over the world and a calculation of how much of each element was contained in each sample. Such an investigation could not provide exact results, but Clarke and Washington's figures show the relative amounts of the elements; *Fig. 1* illustrates this analysis of the earth's crust.

Oxygen and silicon, which are present in granites, sandstones, and most rocks, make up the major part of the composition of the earth's crust, the elements, of course, being in chemical combination with others. Compounds of aluminium, iron and magnesium also exist in large quantities. Some comparatively unfamiliar metals occur in considerable amounts; for example, titanium makes up one part in 160. On the other hand copper, tin, nickel, zinc, lead, mercury, silver and gold, which are essential to our needs and civilization, are among some of the rarest elements in the earth's crust. Thus copper is present to the extent of only one part in ten thousand and lead only one part in fifty thousand.

This fascinating statistical exercise has been attempted many times since 1927. American, German and Chinese geologists, with the increased

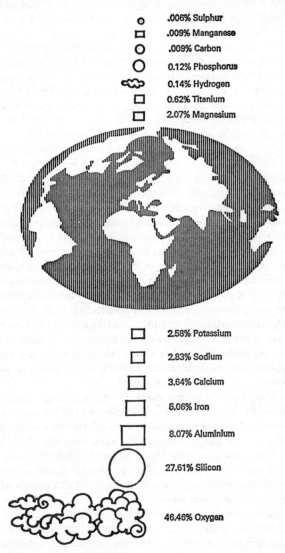

.006% Sulphur
.009% Manganese
.009% Carbon
0.12% Phosphorus
0.14% Hydrogen
0.62% Titanium
2.07% Magnesium

2.58% Potassium
2.83% Sodium
3.64% Calcium
5.06% Iron
8.07% Aluminium
27.61% Silicon
46.46% Oxygen

Fig. 1. Analysis of the earth's crust

knowledge that has been accumulated during the last fifty years, have produced figures remarkably close to those obtained by Clarke and Washington. For example, Professor K. H. Wedepohl gave oxygen as 47·25 per cent, silicon as 30·54 per cent, aluminium as 7·83 per cent and iron as 3·54 per cent.

The availability of metals does not depend on the amount present in the earth, but on the ease with which their ores can be obtained and smelted. An ore deposit near Chicago or Kiev is likely to be more useful than similar ones under the ice of Greenland or at the bottom of the ocean under the Mindanao Trench. The very large iron ore reserves in Brazil are estimated to be 50 000 million tonnes but, because of their remoteness, these have begun to be exploited only recently.

A customer of the local inn would doubtless agree that half a pint in one glass is preferable to sixty teaspoonfuls in sixty glasses, and an analogy can be made with regard to the use of metallic ores. A single rich, extensive ore deposit of a comparatively scarce metal can be exploited more readily than many small pockets spread all over the globe. Ores of copper, lead, zinc and nickel form only a small proportion of the earth's crust, but they are often found in deposits which can be mined in sufficient quantities to supply man's present needs.

Although one twelfth of the earth's crust by weight consists of aluminium, many compounds of this metal are unsuitable to be used as ores. Clay contains about 25 per cent aluminium in chemical combination with silicon, oxygen, iron, calcium and magnesium. A hundred barrowloads would contain enough aluminium to make a small aeroplane, but so long as rich deposits of bauxite, the principal ore of aluminium, are available in Australia, Guinea, Jamaica and elsewhere, the expensive refining processes to extract aluminium from clay will not be required.

The locations of ore deposits, called placements, are distributed somewhat irregularly over the earth. There are some places where Nature has been lavish, as for example in the north of Western Australia, where an area known as the Hammersley Iron Province contains massive deposits of very high grade iron ore. In South Australia there are older 'iron mountain' deposits, with splendid names such as 'Iron Knob', 'Iron Baron' and 'Iron Prince'. Middle East countries are less well endowed with metals, though the salt of the Dead Sea may be regarded as a possible source of magnesium.

Figs. 2, 3 and *4* attempt to give 'score cards' of the countries where ore deposits of various metals are mined; first those major metals which are discussed in Chapters 10–18; then the most important among what we call minor metals in Chapter 19. The tonnages are expressed in terms of

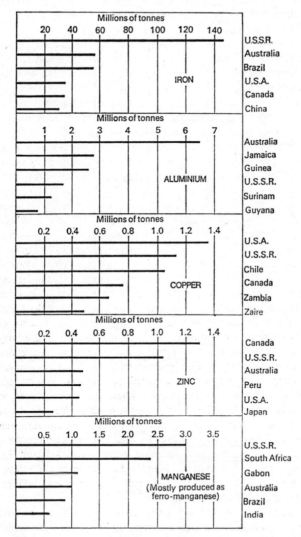

Fig. 2. The first six countries which produce the most ore for some important metals (*The tonnages represent metal content of the ores*)

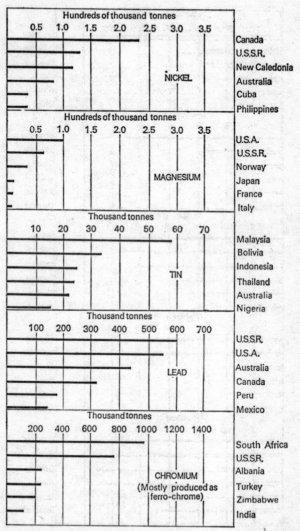

Fig. 3. The first six countries which produce the most ore for some other important metals (*The tonnages represent metal content of the ores*)

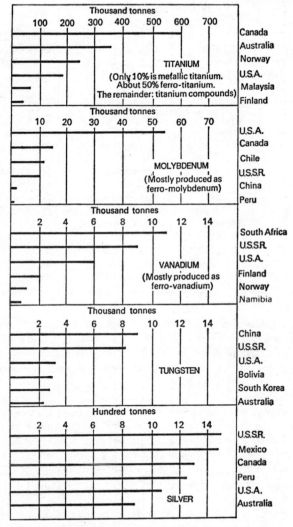

Fig. 4. The first six countries which produce the most ore for some minor metals (*The tonnages represent metal content of the ore*)

Table 2. Some minerals from which metals are obtained

Metal	Name of mineral contained in the ore	Metallic compound contained in mineral	
		Chemical name	Chemical symbols
Aluminium	Bauxite	Hydrated aluminium oxide	$Al_2O_33H_2O$
Copper	Copper pyrites	Copper–iron sulphide	$CuFeS_2$
Iron	Haematite	Iron oxide	Fe_2O_3
	Magnetite	Iron oxide	Fe_3O_4
Lead	Galena	Lead sulphide	PbS
Magnesium	Magnesite	Magnesium carbonate	$MgCO_3$
	Dolomite	Magnesium–calcium carbonate	$MgCO_3CaCo_3$
Mercury	Cinnabar	Mercury sulphide	HgS
Nickel	Pentlandite	Nickel–iron sulphide	$(Fe,Ni)_9S_8$
Silver	Argentite	Silver sulphide	Ag_2S
Tin	Cassiterite	Tin oxide	SnO_2
Titanium	Rutile	Titanium oxide	TiO_2
Zinc	Zinc blende	Zinc sulphide	ZnS
	Calamine	Zinc carbonate	$ZnCO_3$

the metal contents of the ores – even in the case of titanium, of which the greatest amount is used as chemical compounds in, for example, the manufacture of paints. It will be observed that metals such as manganese and chromium, which are associated with the steel industry, are produced in very large quantities.

The majority of useful metallic ores contain the metal combined with oxygen, sulphur, or other elements, as shown in *Table 2*. Pure minerals are rarely found in nature; they are generally contaminated with gravel, limestone, sand, clay, and stones. This unwanted material is termed 'gangue' (pronounced to rhyme with 'hang'). Although some ores are comparatively rich, others contain large amounts of gangue; for example, zinc ores have less than 10 per cent metal, nickel ores about 1·5 per cent and copper ores less than one per cent.

The economic value of an ore depends on the price of the metal and its ease of exploitation. An iron ore yielding less than 25 per cent metal is reckoned as poor while one part of gold in 100 000 parts of ore is rich;

Vaal Reefs, a large South African gold mine, treated 11·5 million tonnes of ore between 31 March 1986 and 31 March 1987, to win only 80 tonnes of gold.

OBTAINING THE ORE

Many ores are recovered from the earth by deep mining in which a shaft is sunk and the ore is blasted out in underground galleries and hauled to the surface. Gold mines are by far the deepest. East Rand Proprietary mines have gone down to 3474 metres, while at Western Deep Levels, the world's deepest mine, gold has been removed at 3582 metres below the surface. At such great depths the technical problems and those of human endurance are immense. The rock temperature is around 52°C (125°F) and, to achieve an acceptable working temperature of 27°C, large quantities of chilled water are provided from refrigeration plants situated underground or at the surface and then distributed through a pipe network. Efforts to cool the mines are now being taken a stage further by circulating a slurry of water and ice. In large mines more than 70 000 cubic metres of cold air are required per minute.

When metallic ores are on or near the surface of the earth, they can be removed by quarrying on a scale that makes one realize how much we depend on earth-moving machinery. The Bingham Canyon copper mine, near Salt Lake City, excavates and smelts sufficient ore to make about 180 000 tonnes of copper per annum. Every day about 100 000 tonnes of overburden and ore are removed, the ore containing only about 0·7 per cent copper in the form of copper–iron sulphide plus a great amount of unwanted earthy material. *Plate 1* shows an aerial view of the quarry; the concentric rings are the terraces which have been excavated. The work is done with the aid of twelve huge electric shovels, six drilling machines, several trains and 35 trucks, each with a capacity of about 200 tonnes. Most of the unwanted material is separated from the sulphide mineral, which contains about 28 per cent copper. It leaves the concentration plant in the form of a slurry which flows twenty miles to the smelting plant, where the water is removed and the ore smelted to copper.

Some metallic ores occur as waterborne deposits and methods based on the familiar 'panning' principle of the old-time gold miner are employed. A more up-to-date form of panning is that used in the recovery of tin ores in Malaysia, where huge dredgers dig up the tin-bearing gravel from lake bottoms and pass the wet gravel over long troughs fitted with shallow projections which retain the dense tin ore.

CONCENTRATING THE ORE

Iron ores contain 10 to 30 per cent unwanted material – earth, sand and clay. These contain silica, which when melted becomes a viscous liquid that tends to combine with iron oxide and other metallic compounds. To prevent this from happening limestone is mixed with the iron ore and, in the heat of the blast furnace, it combines with the silica, forming a fluid slag which can be removed separately.

Magnesium and aluminium are made by electrical processes, which will not operate unless a very pure metallic compound is separated first. A chemical method of concentrating aluminium ores is described in Chapter 4. Ores of copper, zinc, tin and lead contain only a small amount of metal, so the unwanted material must be removed before smelting. Some weak ores are concentrated by passing them, with a stream of water, over reciprocating metal tables. This causes the light earthy material to move in one direction and the heavier metallic compounds to be delivered to another outlet.

In contrast with this, the widely used flotation process causes the dense metal compounds to float in a bath of frothed liquid while the unwanted minerals sink. A story is told about the discovery of this process. In a certain leadmining village, a woman was washing her husband's working clothes at a tub in her backyard. The water was full of suds and the clothes were impregnated with particles of galena, the lead ore which he mined. The peculiar thing about this wash-tub was that the soapsud bubbles, instead of appearing as a white foam, were distinctly dark. Some observant fellow noticed that the surface of the suds was covered with tiny particles of galena. This was surprising, for the lead ore, being very dense, would be expected to fall to the bottom of the tub and not be floating on the surface of the water. The anonymous observer saw in this wash-tub the germ of a commercial process for separating finely crushed lead ore from the gangue associated with it. Americans who tell this story maintain that it occurred in an American mining village; British narrators insist that it was in Derbyshire.

The finely crushed ore is agitated in water containing one or more chemical 'frothing' reagents. The earthy matter is usually 'wetted' and sinks, but the metal-containing particles rise in the froth, whence they are skimmed off and dried. The development of flotation ranks in importance with the discovery of smelting. Virtually the entire world's supplies of copper, lead, zinc and silver ores are first collected in the froth of the flotation process.

LEACHING THE ORES

Pellets or nuggets of gold were discovered and used earlier than 4000 B.C. when powerful monarchs, like those in Egypt, could afford the manpower to separate grains of gold from alluvial gravels, though these dust-like particles were not easily consolidated by melting. Sands containing grains of gold were washed with sheepskins and the metal was retained in the hairs, the lighter sand being washed away. It is believed that when Jason and his Argonauts went seeking the Golden Fleece, their voyage round the Black Sea was in search of gold. In those days the Colchian peoples at the eastern end of the Black Sea collected gold dust on the fleece of sheep, as it was washed down the river Plasis in the land of Colchis. This ancient process has been superseded by treating gold ores with potassium cyanide. The development of a method of producing this chemical cheaply and of utilizing it in the extraction of gold was a momentous discovery which greatly influenced the development of gold mines and of international trade in various parts of the world.

After crushing to a fine powder, the ore is agitated with a very dilute solution of cyanide in large tanks, each holding about a thousand tonnes of crushed ore and solution. Once the gold has been dissolved by the cyanide, the solution is filtered and the gold precipitated from the solution. *Plate 2* shows an aerial photograph of West Driefontein, one of the largest South African gold mines. The settling tanks are in the bottom left-hand corner. The cyanide plant is just below the centre of the photograph, comprising 18 tall tanks known as Pachucas.

FROM ORE TO METAL

As was seen from *Table 2*, several metallic ores contain the metal combined with oxygen; for example high-grade iron ores contain iron oxide. This metal is cheap because its ores contain a large proportion of iron oxide which can be smelted on a big scale in blast furnaces, with the expenditure of comparatively little fuel, time or labour per tonne of iron. On the other hand, although the aluminium ore, bauxite, contains aluminium oxide, it cannot be changed into metal by smelting in a blast furnace. A large amount of electrical energy is needed to separate aluminium from the oxygen with which it is chemically combined.

As supplies of rich metallic ores cannot be obtained for much longer, worldwide research has been devoted to the extraction of metals from weak ores or even from the depleted tailings dumped as residues from previous extraction processes. During the last twenty years a process known as solvent extraction has advanced from an expensive exercise to a metallurgical process that is coming to be of immense importance. Solvent extraction depends on a complex series of reactions in which organic chemicals dissolve the metallic compound, which is then separated; then further processes reverse the reaction so that the organic solvent is reclaimed and the liquid containing the metal goes on for electrolytic extraction. More than forty proprietary solvents are available and of these about twelve are in everyday use.

At first this process was used for separation of the less common metals such as zirconium, hafnium, niobium, uranium and tantalum. It was also a satisfactory method of separating metals of similar properties – such as niobium from tantalum, cobalt from nickel and the various 'rare earth' metals listed on page 218. Then the pace of development quickened and it has been predicted that by the end of the century the revolutionary impact of solvent extraction will be comparable with that of froth flotation at the beginning of the century.

Nowadays the most exciting and challenging development of solvent extraction has been in the separation of copper from mine dump leach liquors, which previously had been uneconomic or impossible as sources of copper. Several plants producing over ten tonnes of copper per day are operating in Arizona, Montana, Texas and Nevada, while one in Zambia has a potential of over two hundred tonnes per day. A still larger plant was designed for that much-troubled country, Zaire, but lack of finance resulted in only a small operation being planned, to extract copper and the expensive metal cobalt. Solvent extraction has encountered many problems and it is still admitted that the purity of the copper resulting from the process leaves something to be desired.

The science of biotechnology is opening new fields for metal extraction in the future. Metal sulphide minerals are leached and interacted with bacteria which convert the sulphides into oxides. Often this process is followed by solvent extraction and finally the metal is extracted from the enriched ore. This method of metal recovery using micro-organisms is showing promise for the treatment of weak ores of copper, nickel, cobalt and uranium.

The smelting of the two best-known metals will be described in the following chapters, first iron and then aluminium. The production of zinc is mentioned on pages 181–2, nickel on page 189, magnesium on page 200 and titanium on page 201.

3

MAKING IRON

When an ore containing iron oxide is heated with carbon, metallic iron is produced. The chemical processes taking place in the blast furnace are many and complex but, bringing them to the simplest form:

OXIDE OF + CARBON → IRON + OXIDE OF
IRON   CARBON
(ore)   (coke)   (metal)   (gas)

   This is the fundamental process on which iron smelting is based, though other associated reactions take place at the same time. In addition secondary chemical changes convert the earthy non-metallic materials, contained in the ore, into slag.

   The production of iron began in the second millennium B.C. and the metal was being produced on quite a large scale by 1200–1000 B.C. In those early times iron ore was heated in a charcoal fire (doubtless by chance at first). When the fire had died down a spongy mass of iron remained which could be hammered into shapes and used for tools and weapons. Our metallurgical forefathers found that when they blew or fanned the flames the fire became hotter and the iron was produced more rapidly, so bellows were used to increase the supply of air. Even so, the heat of the ancient furnaces was not sufficient to reach the melting point of iron, about 1535°C.

   In Britain during the fifteenth century, furnaces of sufficient size with a forced blast of air were used to produce a form of iron which could be poured from the furnace. This metal contained 3–5 per cent of carbon and some other elements such as silicon and the cast iron melted at only about 1200°C, a temperature that could be obtained in those early blast furnaces. It must have been surprising to discover that cast iron, or 'pig iron', was brittle whereas the forged iron, which had not been melted, could be shaped by hammering. Nevertheless cast iron had the advantage that it could be poured into moulds and formed into the required shape. The forgeable, nearly pure, iron from the charcoal fire did not absorb much carbon but when iron ore was smelted in a blast furnace, carbon was absorbed.

At first, charcoal obtained from wood was used for smelting iron. One furnace, built in 1711 at Backbarrow near Lake Windermere, was 6 metres high and 2½ metres square on the outside; it could produce 6 tons of iron per day. Typically, it was built into the side of a hill so that the workers could tip their baskets of ore and fuel into the furnace from a platform on the hillside. The work was done by seven men, aided by their wives and families.

Early in the twentieth century, the square stone stack, which had been rebuilt in 1870 to contain a circular hearth, was still being fuelled by charcoal produced in the local woods, but production ceased during the 1914–18 war. In 1926 the furnace was converted to the use of coke and continued in operation until the early nineteen-sixties. Today the works, not far from the Dolly-Blue factory, lie derelict. However, of the seven charcoal furnaces that worked in south Lakeland, Duddon Furnace (1736–1867) is being restored by the Department of the Environment. Also a 'finery-chafery forge', at Stoney Hazel, about three miles north of Backbarrow, is being reconstructed. These two sites will eventually represent complete examples of the charcoal iron furnace industry in England. The Bonawe furnace in Scotland and the Dyfi 'Ffwrnais' at Eglwys-fach on the road between Machynlleth and Aberystwyth are tourist attractions. The Welsh example has a splendid reconstructed water-wheel similar to the one which provided the blast.

When charcoal was used to smelt iron, about one acre (4000 square metres) of woodland had to be cut down per ton of iron.* The wholesale destruction of forests for this purpose made it necessary to experiment with other materials. In 1708, Abraham Darby, a young man who was apprenticed in Birmingham and then managed a brass foundry in Bristol, leased a disused 160-year-old blast furnace at Coalbrookdale. A year later he succeeded in smelting iron with coke. After his death his efforts were continued by his partners, then by his young son and later by his grandson. The improvements and cost savings of Darby's achievement provided the starting point for the manufacture of iron rails, the iron bridge over the Severn, iron boats, iron aqueducts, iron buildings and eventually the worldwide ferrous industry.

* It is deplorable that thousands of square miles of Amazon forest are threatened with destruction because Brazilian charcoal-burning furnaces are already making over 100 000 tonnes of pig iron per year for export to Europe, Japan and China.

IRON ORES

Iron forms several oxides, which are available as iron ores. Haematite contains Fe_2O_3 which, if pure, would give 70 per cent iron. Magnetite is a magnetic iron ore, the oxide being Fe_3O_4. There are other ores containing iron carbonate.

The world's annual production of iron is more than 600 million tonnes, for which 1000 million tonnes of iron ore is required. Four regions, the U.S.A., the U.S.S.R., Japan and Western Europe, each make about 100 million tonnes of iron per annum and need enough ore to manufacture that amount of metal. The U.S.A. provide two thirds of their own ore and import one third, about half of which comes from Canada, which is a large exporter. The U.S.S.R. and Eastern Europe get most of their supplies, appropriately, from inside the Iron Curtain. The U.S.S.R. is the biggest iron ore-producing country. To the east of Kiev a vast deposit of magnetic iron ore, known as the Kursk Magnetic Anomaly, contains an estimated 80 000 million tonnes of rich iron ore providing 45–65 per cent metal content, plus about 10 million million tonnes of poorer ore. This area covers some 200 000 sq. km and is said to contain enough ore to produce 250 million tonnes of iron a year for the next 15 000 years.

Japan is a large importer of iron ore, able to bargain keenly, offering long-term contracts with Brazil, India, South Africa, New Zealand and other countries. Europe contains several major iron and steel producers: Britain, West Germany, Italy and France. The deposits of native ores in Lincolnshire and Northamptonshire used to provide most of Britain's requirements but they contain only about 30 per cent of iron. Recent technical developments have called for iron ore of far higher grade than the U.K. can supply and we now import most of our ore, averaging 62 per cent iron content, from Canada, Australia, Venezuela, Liberia, Mauritania, Brazil, Spain, South Africa and Sweden.

Today, the sea transport of iron ore represents the largest dry bulk seaborne commodity in the world. The two next largest, grain and coal, are together less than half the weight of the iron ore trade. Economic transport requires large vessels; in 1977 the average size of ore carriers was about 120 000 tonnes, but now there are several of more than 200 000 tonnes capacity with some carriers in the region of 300 000 tonnes. The new breed of ore carriers has made it viable to transport ore from many parts of the world. To permit the use of such vessels, deep ore terminals have been constructed for loading and unloading.

New Zealand was responsible for the Marcona process, an impressive development in the treatment and transport of iron ores. The west coast of North Island contains massive deposits of black ironsands, which are mined and concentrated at Taharoa, 150 km south of Auckland, for export to the Japanese iron and steel industry. Using gravity cone separators and then wet magnetic drums the oxide is separated from extraneous material and pumped in slurry form through pipelines to bulk ore carriers moored offshore. The highly mechanized slurry-loading technique is the most advanced of its kind in the world. In addition, New Zealand Steel's Glenbrook plant, 50 km south of Auckland, utilizes ironsand concentrate from nearby Waikato North Head in the manufacture of iron and steel. New Zealand Steel's unique direct reduction process is the first in the world to use ironsand and non-coking coal in full-scale operations.

THE BLAST FURNACE

Fig. 5 shows the basic features of a modern blast furnace – one of the most advanced in the world: the No. 1 furnace at British Steel's Redcar site. The furnace will be described later, on page 28.

Most iron ores contain 60 to 80 per cent iron oxide, mixed with sand, earth, clay and stones which contain silica and other compounds. If such a mixture were smelted in a blast furnace, a great deal of the iron would combine chemically with the silica; this was what happened in primitive iron-smelting furnaces. It is therefore necessary to introduce another material which will combine with silica, thus preventing it from uniting wastefully with the iron. Limestone has this property, and at the temperature of the blast furnace it forms chemical compounds with the earthy materials including the silica, the result being known as slag. This subsides down the blast furnace; being lighter than iron, it then floats above the layer of molten metal which is accumulating at the bottom. The two molten materials, iron and slag, can then be removed separately at two different levels.

A blast of hot air, enriched with some additions, is blown through a number of nozzles called tuyères. Depending on the capacity of the furnace there are between 12 and 36 of these, distributed evenly around the circumference and placed sufficiently high above the hearth to leave a space in which the molten iron and slag can accumulate. Steam or oxygen may be added to the blast; several forms of hydrocarbon are also injected at the tuyères – oil, natural gas and powdered coal, though not all furnaces are equipped for all these options. The main attraction of enrichment is that the volume of the 'burden' can be reduced so that more iron can be made in a furnace of given size.

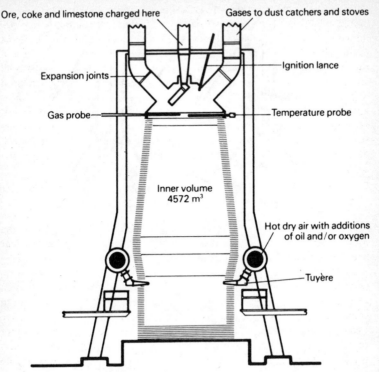

Fig. 5. British Steel's 10 000 tonne per day blast furnace at Redcar

CHARGING THE FURNACE

The ore, coke and limestone are given the collective name of the 'charge'. The method of feeding the charge into the furnace is somewhat complicated, the object being to avoid wasting any of the blast-furnace gas and to distribute the contents of the furnace evenly. The charge is hoisted to the top of the furnace and in some plants it is dumped on to a double bell and hopper arrangement, shown in the left-hand drawing in *Fig. 6.* The small bell is then lowered, as in the right-hand drawing, dropping the charge into a lower chamber, at the bottom of which is a larger bell and hopper. When the top has been closed, thus sealing the furnace, the larger bell is lowered and allows the charge to fall into the furnace. The bell is rotated through an angle every time it is used.

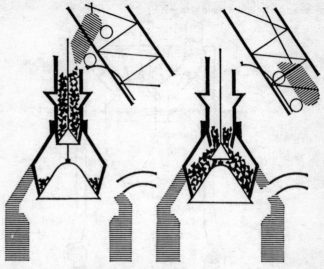

Fig. 6. Bell and hopper for charging a blast furnace

A number of furnaces are now employing other methods of charging, such as belt delivery to the top of the furnace, to provide better sealing and distribution. Very large furnaces introduce serious problems in the handling of bell and hopper charging-equipment; for example, the large diameter top ring associated with bell charged furnaces is difficult to maintain pressure-tight and under the high pressures prevailing in modern, large furnaces, the problem becomes acute. A recent development to overcome these difficulties is the bell-less top, in which the charge is distributed from sealed bunkers at the furnace top, by means of a rotating chute. *Fig. 7* shows the bell-less top installed at the No. 1 Redcar furnace and two at Ravenscraig.

THE HOT BLAST

The primary function of the air blast is to enable the coke to burn and produce a high temperature. It also combines with the burning coke to form carbon monoxide which itself reacts with the iron ore, producing iron and carbon dioxide.

Until about a hundred and fifty years ago, cold air was blown into all furnaces; indeed until the 1930s there were several 'cold blast' furnaces,

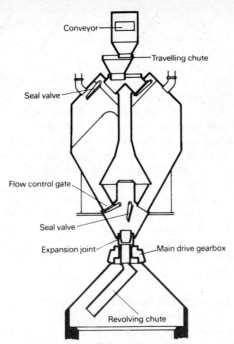

Fig. 7. Bell-less top for charging a large blast furnace

making pig iron. When a cold blast of air was used, as much as eight tons of coal or coke were needed to produce one ton of pig iron. In 1828, a Scot, James Neilson, developed the use of a heated blast and the fuel consumption was reduced to five tons per ton of iron, a great achievement at the time, though the efficiency of modern blast furnaces is about ten times better than that of Neilson. Those who wish to pay tribute to this great metallurgist of the past can find an obelisk to his memory on a hilltop near Kirkcudbright.

Fifty years ago the hot gas was allowed to burn at the top of the furnace, but now it is not wasted in this manner; it provides an excellent method of pre-heating the blast. On leaving the furnace this gas contains 20 to 25 per cent carbon monoxide, which is combustible. Some of the furnace gas is used for heating the air blast by means of 'stoves', usually three to each furnace. These stoves are nearly as large as the blast furnace itself; they consist of, firstly, a chamber in which the remaining carbon monoxide is burnt and, secondly, a labyrinth of firebricks, arranged in such a way that the burning gases in passing through them heat the firebricks, making them red hot. In operation, two processes are carried out at the same time in the stoves:

1 two stoves are being heated, as described above;
2 the other stove, having been heated, raises the temperature of the blast of air as it passes to the furnace.

By this method, the air is at a temperature between 1000 and 1350°C when it reaches the blast furnace. About every half hour these operations are changed; the stove which previously was 'on blast' is heated, while one of the stoves which were 'on gas' now gives up its heat to the incoming blast of air. Normally the cycle is half an hour on blast and one hour on gas. As will be seen from *Fig. 8*, the gas is passed through dust extractors before being led to the stoves. The remainder of the valuable furnace gas is used for other purposes – for driving pumping-engines and for making sufficient electricity to provide light in the works and often for a neighbouring town as well.

SINTERING

In order to achieve maximum production the ore must first be prepared. This is usually done by crushing and screening the ore and then roasting it with limestone and with coke in the form of fine particles. In this process of agglomeration, the volatile matter is driven off by the heat; the limestone flux and the coke are incorporated with the ore, and the agglomerate is produced in the form of irregular clinkery lumps, known as 'sinter'. Agglomeration produces a strong material which can be smelted efficiently in large blast furnaces. There is another good reason for sintering. Ore particles of less than 5 mm size would be blown out of the furnace by the strong air blast and so, to make full use of the ore, the 'fines' are agglomerated in the sinter and thus are reduced to iron in the furnace instead of being blown away in the exhaust gases.

In 1960 about 40 per cent of the burdens of British blast furnaces consisted of sinter; by 1974 the proportion had risen to 58 per cent and by 1987 to about 80 per cent. Ores too fine for sintering are formed into pellets, usually at the ore fields. These resist damage during transportation and make excellent feed for the blast furnace. In recent years pellets have become widely available on the world market, though in Britain the one plant at Redcar was under-utilized and is now closed.

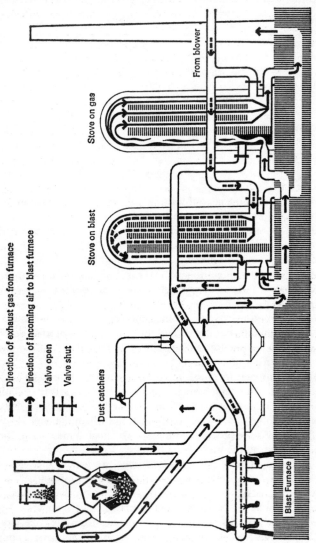

Direction of exhaust gas from furnace

Direction of incoming air to blast furnace

Valve open

Valve shut

Stove on gas

Stove on blast

From blower

Dust catchers

Blast Furnace

Fig. 8

Table 3. Tonnages to be handled daily by Europe's largest blast furnace

	Input tonnes		Output tonnes
Sintered ore	11 200	Iron	10 000
Rubble ore	4 800	Slag	3 000
Hot-blast air	14 000	Furnace gas	22 000
Oil in hot blast	300		
Coke	4 250		
Limestone	100		
Total	34 650	Total	35 000

THE IRON-MAKING PLANT

The furnace itself is the central feature of a complex organization for the production of iron and, although by its massiveness it dominates the scene, it is only one link in a chain of processes. Plant must be provided to carbonize coal to coke; to pump and heat the air blast; to receive, store and charge into the furnace the coke, sinter, and limestone flux; to clean and handle the vast quantities of the gas which discharges from the furnace; and to handle and prepare for sale or disposal the slag which is produced.

Table 3 is not quite correct arithmetically, because the production of a furnace is continuous, but it indicates the tonnages of materials that have to be handled when the Redcar No. 1 furnace produces 10 000 tonnes of iron per 24 hours.

The flame temperature at the tuyères is about 2200°C and the contents of the furnace reach a temperature of about 1800°C, nearly 300°C higher than the melting point of pure iron and about 550°C higher than that of the iron containing about 4 per cent of carbon which is being produced in the furnace. As the process continues, iron is being made from the ore and then melted by the heat of the furnace. Words like 'trickle' or 'fall' are inadequate to describe the conditions in which the enormous contents of the furnace move downwards at the rate of 500 tonnes per hour, while the blast of air rushes upwards. Finally the hot metal and slag accumulate at the bottom of the furnace in two layers. During its descent the iron absorbs a number of impurities from the coke, ore and limestone, so that the pool of iron at the bottom of the furnace contains about 3 to 5 per cent carbon, about 1 per cent manganese, up to 3 per cent silicon, and usuallY a small amount of sulphur. In some types of iron, phosphorus is present up to 2 per cent; in others the phosphorus content is only a fraction of 1 per cent.

When the level of the iron and slag has risen to approach the tuyères, the slag is drawn off and the metal is 'tapped'. For this blast furnaces are provided with two openings or 'tap holes' which are sealed with plugs of refractory material, the upper opening for slag and the lower one for molten iron. When the time arrives to remove the molten pig iron, the lower tap hole is broken open with a thermal lance and the white-hot liquid metal gushes out. Sometimes it is run into moulds, forming ingots of pig iron, but often there is a steelworks near the blast furnace and the iron is conveyed molten to be converted into steel.

The slag, which is run off separately from the iron, can be regarded as a calcium–aluminium silicate, containing several other compounds. According to the types of ores used and the grade of iron manufactured, slags with different physical and chemical properties are produced. They find various uses such as railway ballast, tarmacadam, concrete, porous slag bricks for building, fertilizers, and slag wool for heat insulation. These do not absorb all the available slag and some of it has to be dumped.

MODERN BLAST FURNACES

From the small furnaces of several hundred years ago, the modern blast furnace, with all its ancillary plant, has developed. One of the most impressive achievements in metallurgy has been the growth in size and increase of efficiency of the iron blast furnace, comparable with the difference between the *Mayflower* of 180 tons and the *Queen Elizabeth 2* of 65 000 tonnes.*

A typical output of a blast furnace in the 1780s was 900 tons per annum but now some very large furnaces make that amount in two hours! Nippon Steel's No. 5 blast furnace at Kimitsu broke the world record in April 1980 with a daily average of 11 670 tonnes, achieving this enormous production with the help of oxygen injected into the air blast. Another highly productive Japanese blast furnace is illustrated in *Plate 3*. The construction of a furnace of that size requires 4500 tonnes of steel for the furnace body and another 4000 tonnes for piping and machinery. About 6000 tonnes of refractory blocks of various qualities and sizes, mostly over a tonne in weight, are required in Japan, where earthquakes

* See page xvi for a comment about tonnes versus tons.

have to be anticipated. The foundations of such a monster contain 10 000 cubic metres of concrete. The steel shell and supporting columns are then installed and assembled; the furnace lining follows, to provide a bottom more than 3 metres thick and the furnace walls over a metre thick. There is a one-month drying period and the whole building programme takes about 20 months.

Like a motor-car, but much more so, a blast furnace needs long and careful running-in. Before the blast begins in Japanese furnaces about 18 000 old wooden sleepers are stacked inside, then charcoal and firewood are added, after which coke and iron ore are placed in alternate layers. When this has burnt for 24 hours about 500 tonnes of pig iron will have been produced, the first of over 20 million tonnes which the furnace will make in its lifetime.

Japan's development of very large blast furnaces illustrated a principle that was so logical that other iron-producing countries probably wished they had realized it before. An important economic factor in running a blast furnace, and indeed any process, is to achieve maximum output with minimum breakdown time. The really costly breakdown in a blast furnace involves relining, which is made necessary by the attack of the molten cascade of iron and slag descending the furnace and the red-hot air blast hurricaning up it. This is related to the surface area of the furnace lining; but quadrupling the volume of any container involves only doubling the inner surface area, so the bigger blast furnace will be expected to give less relining cost per tonne of iron produced than a smaller furnace.

During the recession in the late 1970s, British production of iron declined and the efficiency of the plants compared unfavourably with those of other countries. Then, in preparation for better years ahead, many impressive changes were introduced. British Steel developed what might be called Jumbo blast furnaces, the largest of which, at Redcar on the south bank of the Tees, has a capacity of about 10 000 tonnes per day. A diagram of this furnace is shown on page 23.

Size alone was not the only development. Every aspect of iron production has been, and still is being, attacked to bring Britain securely to the top of the 'first division'. An immense blast furnace would be vastly unprofitable unless its production was supported by a wide range of controls. For example, the four furnaces at Scunthorpe (named after the Queens Mary, Bess, Anne and Victoria) inject granulated coal into the blast and monitor each phase of production with computers. They use what is known as movable throat armour to deflect the materials as they fall off the large bell and to give maximum efficiency of the flow of the blast, so that it is at a minimum near the refractory walls of the furnace, thus reducing downtime.

During the mid-1970s, the fuel consumption of blast furnaces worldwide was reduced from about 650 kg of coke per tonne of iron to about 500 kg per tonne, with Japan in the lead. Now the more efficient British furnaces are obtaining still better fuel efficiencies. When the Redcar furnace is fuelled by coke alone it requires 465 kg per tonne of iron. When 30 kg of oil per tonne of coke is injected into the blast, the coke consumption is down to 425 kg per tonne.

NEW DEVELOPMENTS IN IRON-MAKING

Every industry, large or small, needs to take a look at itself fairly frequently. Labour and fuel costs increase, raw materials become scarce as supplies are exhausted or become plentiful due to new finds; capital costs mount; new processes are discovered. The trouble very often is that, soon after an industry has invested a great deal of money in capital equipment, some new factors arise. This is particularly true in the iron and steel industry. If you have just spent £500 million on a new iron and steel works the prospects of scrapping it are painful. Blast furnaces have undoubtedly increased in efficiency but coke, on which blast furnaces have depended for over 250 years, is becoming more expensive every year, and supplies of suitable coking coals are becoming scarcer.

Many iron-makers consider that the time has come to reduce iron from the ore in a form that can be converted directly into steel. This was begun experimentally in the early 1950s but was not pursued energetically because the improvement of blast furnace efficiency was coming into its stride, so direct reduction process developments were put off till later by the big industrialized countries.

In Mexico an increasing demand for steel had made it necessary to import scrap, with the resultant problems of price fluctuations and unreliable quality. In response, the steelmakers HYLSA pioneered a revolutionary direct reduction process, which enabled them to use local ores and fuel. Natural gas is mixed with superheated steam and converted into carbon monoxide and hydrogen. Pellets of iron ore are treated with the gaseous mixture at high pressure in a furnace, at a temperature which is high but not sufficient to melt the iron. In this way the ore is converted to spongy iron pellets of a high standard of purity, suitable for conversion to steel.

During the past decade there has been continuous development of the process and new plants built, though the works in Monterrey which began production in 1957 is still making about 1·0 million tonnes per annum. Another in the same district which began in 1960 has been converted to a more modern design. One Mexican plant in Las Truchas has a capacity of 2·0 million tonnes per annum. Other established plants, ranging in capacity from about 250 000 to 1·1 million tonnes per annum, are in Brazil, Venezuela, Indonesia, Iraq and Iran. About twelve million tonnes of iron per annum are produced worldwide by the H Y L process.

In another new process which is being used in Canada and Nigeria, ore containing the magnetic iron oxide Fe_3O_4 is crushed and the unwanted silica washed away. Then, by magnetic methods, almost pure oxide with an iron content of about 68 per cent is separated and is suitable for direct reduction. There is a plant at Contrecoeur, near Montreal, where the iron oxide is treated in a vertical shaft furnace by a reducing gas generated on site. This is known as the Midrex process. The iron pellets so produced are cooled at the base of the furnace and transferred to an electric arc furnace for melting and making into steel. The remainder of the iron ore is suitable for separate smelting in blast furnaces.

At present there are at least fifty other processes for the direct reduction of ore to metal; most of them have been only for scientific interest and only a few of them are working on a commercial basis. Some direct reduction processes are economic even in comparatively small units, and they often link up with the production of medium-sized or mini-steelworks operated by countries which are not yet highly industrialized and which therefore do not possess the vast amount of scrap iron and steel on which the major iron- and steel-making countries depend.

Price rises of natural gas have undermined the economic viability of direct reduction processes so they have continued to operate mainly in countries which have ample supplies of natural gas. In Europe, the U.S.A. and Japan, several pilot plants have been built to produce iron by smelting with coal instead of coke. This has the advantage that expensive coking-ovens are not required. So far it has been found that coal-based ironworks will be economic where old plants are being completely rebuilt or where capacity in an existing blast furnace plant is being increased. At present there is over-capacity in the western world and it is likely that these developments will take place only slowly, though the techniques of using coal will continue to be studied closely.

4

MAKING ALUMINIUM

Aluminium is the most plentiful metal in the earth's crust, of which it forms 8 per cent. The pure metal does not occur in nature but alum, which is a hydrated potassium–aluminium sulphate, was used in the sixth century B.C. as a mordant to 'fix' dyes in fabric. Other compounds of aluminium were known and used for many centuries before the metal was isolated. It had been given the name aluminium in advance of its discovery in 1824, when Hans Christian Oersted succeeded in making aluminium chloride and then treated this compound with an amalgam of potassium in mercury, producing a lump of white metal which he believed to be an impure form of aluminium but which was probably an amalgam of aluminium in mercury.

Two years later Friedrich Wöhler treated aluminium chloride with potassium metal, forming a grey powder consisting of aluminium heavily contaminated with oxide. Over the next eighteen years the process was improved till pieces of aluminium as big as pin heads were obtained – sufficiently large to measure some of the properties of the metal and to discover that it had a much lower density than the other metals known at that time. A further development came in 1854 when Etienne Henri St Clair Deville used sodium instead of potassium for chemical reaction with sodium–aluminium chloride and made a small ingot of aluminium, but the price of the metal remained high and the output small.

Fortunately the Emperor Napoleon III was seeking a light metal for French armour and he sponsored Deville's researches. When sufficient metal became available Napoleon proudly displayed aluminium cutlery at his State banquets and an aluminium rattle was made for his infant son. At that time aluminium cost about £200 per kg, though as production improved the price gradually lowered to about £4 per kg in 1885.

Deville and others had realized that the metal might be obtained by electrolysis of the oxide, alumina, but although some aluminium was made in that way, only small amounts of electricity could be generated and it was not till after Zenobe Gramme's invention of the dynamo in 1870 that the electrolytic process became feasible. Aluminium oxide possesses a high melting point of about 2000°C so alone it is not suitable for separation into aluminium and oxygen by electrolysis. However, the process could be achieved if a substance were found that would dissolve alumina, forming a solution with a much lower melting point than that of the oxide alone.

These were some of the problems to be solved when, in the 1880s, Charles Martin Hall, in America, and Paul Louis Toussaint Héroult, in France, were experimenting with the electrolytic method, taking advantage of the great new possibilities of cheap and plentiful electric power since the development of the dynamo. They worked separately, each unaware of the other's researches, till in 1886, after heart-breaking failures and little encouragement the two pioneers* both arrived at the same solution by using the mineral cryolite, whose chemical composition is sodium–aluminium fluoride. Hall and Héroult found that cryolite dissolves about 5 per cent of alumina; the solution melts at a little under 1000°C and an electric current passed through it will keep the liquid molten and simultaneously split up the aluminium oxide into aluminium and oxygen.

Hall and Héroult's process soon caused the price of aluminium to fall steeply but at first there was little interest in the metal and the first plant using Héroult's patent obtained the aluminium to make the copper alloy, aluminium bronze. It was not till 1888 that the first aluminium-producing companies were formed, one each in France, the U.S.A. and Switzerland.

BAUXITE

The principal ore from which aluminium can be extracted economically is bauxite, which consists of aluminium oxide combined with water, with the chemical composition $Al_2O_33H_2O$ (aluminium trihydrate). The ore contains impurities in the form of silica (sand) and oxides of iron and titanium; the iron oxide gives bauxite a red colour. Bauxite is named from the district of Les Baux, near Arles in the south of France, where this important ore was first worked commercially. There are valuable deposits in Australia, Brazil, Jamaica, the Republic of Guinea, the U.S.S.R., Surinam and Guyana.

* Hall and Héroult were both born in 1863. Both died in 1914.

Jamaica had mined practically no bauxite before 1952, but by the following year over a million tonnes were obtained, rising to 16 million tonnes in 1976. Then several political and economic problems caused a decline in output to about 6 million tonnes in 1986, since when there has been a gradual increase.

During the past few years Australia has become an Aladdin's cave of almost every mineral that matters; the continent has been called the biggest quarry in the world. The 400 km long Darling Range in Western Australia, Weipa in the Cape York Peninsula and the Gove in the Northern Territory contain bauxite deposits totalling 3500 million tonnes, ten times those of Jamaica.

FROM BAUXITE TO ALUMINA

Bauxite deposits generally lie near the surface; most are located near the Equator in areas where hot sun and heavy rain have weathered the ore over millions of years. After the vegetation and top soil have been removed, the over-burden of sand or clay is stripped away, either by mechanical scrapers or by powerful hydraulic jets, and the bauxite is mined by open-cast methods. Instead of removing the impurities by slagging during smelting, as in the production of iron, the aluminium ore has to be purified before it can be converted into metal. The concentration process, devised by Karl Josef Bayer in 1890, depends on the fact that aluminium trihydrate dissolves in heated caustic soda but the impurities do not; this enables practically pure aluminium oxide (alumina) to be separated.

The ore, containing about 50 per cent of aluminium trihydrate, is crushed and ground to a powder, then mixed at high temperature with a solution of caustic soda (NaOH) in digesters, under pressure. The aluminium trihydrate dissolves in the caustic soda, forming sodium aluminate which, being soluble, can be passed through filters, leaving the insoluble impurities behind. The aluminate solution is pumped into precipitator tanks 25 metres high, in which very fine and pure particles of aluminium trihydrate are added as 'seed'. Under agitation by compressed air and with gradual cooling, pure aluminium trihydrate precipitates on the 'seed' and is then separated from the caustic soda solution by settling and filtering. It is then heated to 1000–1100°C, which drives off the chemically combined water, leaving aluminium oxide as a fine white powder suitable for the electrolytic smelting process. The caustic soda is recovered and returned to the start of the process, to be used over again to treat fresh bauxite.

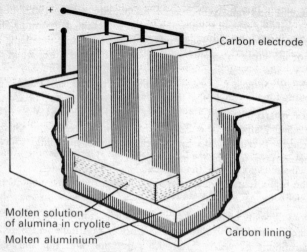

Fig. 9. Extracting aluminium by electrolysis

ALUMINIUM SMELTING

An aluminium reduction furnace is schematically illustrated in *Fig. 9*, though normally it contains about twenty electrodes in two lines, connected to a direct current power supply giving about 200 000 amperes. The furnace is a steel container about ten metres in length, up to three metres wide and 1·5 metres deep, lined with refractory bricks and an inner layer of carbon, which serves as the cathode. The anodes are carbon blocks, each weighing about 750 kg, suspended and dipping into the molten solution of alumina in cryolite. A modern furnace cell produces over a tonne of aluminium every twenty-four hours. Such cells work in series of 150 to 200, called a 'pot line' and rated at about 150 megawatts, using 13–15 kWh per kilogram of aluminium produced. The carbon anodes are made from a mixture of coke and pitch, baked and fitted with the steel rods by which they will be suspended.

Until the beginning of the 1939 war cryolite, mined in Greenland, was used but production of synthetic sodium–aluminium–fluoride had begun and now this chemical has replaced the mineral. The alumina is dissolved in the fluoride in the proportion of about one in twenty; some aluminium fluoride is also added to lower the melting point further.

While aluminium is being produced, the cryolite remains practically unchanged, the furnace being continually replenished with alumina.

Since aluminium is heavier than the molten solution of aluminium oxide in cryolite, it sinks to the bottom of the cell from which it is periodically siphoned into a travelling ladle, conveyed to a large holding furnace and then cast into ingots. The oxygen liberated at the anodes unites with the carbon to form carbon monoxide which burns at the furnace top, making carbon dioxide. The anodes are progressively fed into position; this accounts for the heavy wastage of electrodes, of which about half a tonne is burnt away for each tonne of aluminium produced. The passage of the strong electric current through the cell results in the evolution of heat in much the same way as passing electricity through the elements of an electric fire develops heat. This is sufficient to maintain the minerals in a molten state and to bake the anodes.

The ingredients needed to make a tonne of aluminium illustrate that the smelting of such a metal is a gigantic exercise in mass-handling. Thus $4\frac{1}{2}$ tonnes of bauxite are needed to produce 2 tonnes of alumina by the Bayer process, requiring one tonne of fuel oil and 160 kg of caustic soda. In the electrolysis 15 000 kWh of electricity, 450 kg of pre-baked anodes and 50 kg of cryolite are required.

The economic production of aluminium depends largely on the availability and cost of electric power. Canada, the U.S.A., U.S.S.R., Norway and Switzerland are large producers, because of their low cost hydro-electricity. Britain is short of such power and our aluminium smelting depends mostly on electricity from nuclear energy in Anglesey and a coal-fired power station in Northumberland. The closure of a smelter at Invergordon which had been obtaining power from the national grid emphasized that the price of electricity has a major effect on the cost of producing aluminium. Every one-tenth of a penny in power cost makes a difference of nearly £20 per tonne of primary aluminium.

In recent years some countries in the Middle East and in the Third World have been installing aluminium smelters. Some are built as a kind of status symbol; others, such as in the Gulf States and Egypt, because the countries have cheap surplus energy. In the former cases it is spare natural gas from the oilfields and in the latter hydro-electricity from the Nasser dam on the Nile, near Aswan.

In 1976 Alcoa of America opened a pilot plant in Texas for producing aluminium by a new method. Alumina is mixed with carbon and treated with chlorine in a reactor at 800°C. The resulting gaseous aluminium chloride is filtered, allowed to become liquid in an inert gas atmosphere and converted to aluminium by electrolysis. This process consumes less electricity than Hall–Héroult and does not require the use of cryolite.

However the corrosion caused by chlorine and aluminium chloride led to problems in the plant that were described as a plumber's nightmare. A development by Mitsui Alumina uses coking coal to smelt aluminium from bauxite, but the process requires a temperature of 2000°C.

So far, no new process has been able to match Hall–Héroult, which has made progressive improvements in efficiency. For example there is computer control of the pot lines, with individual microprocessors. The carbon plant is mechanized, the alumina is handled automatically and the anodes are changed automatically.

5

ALLOYS

An automobile contains many different metals and alloys, chosen because of their properties, availability and cost, the duties they will perform and their suitability for the manufacturing processes. About three quarters of the weight consists of ferrous metals, which are iron alloys, including mild steel for bodywork and heat-treated alloy steels for engine components such as gears, connecting rods, camshaft and valves. Cast iron or aluminium alloys are used for the engine cylinder block. The gear box, transmission housing, pistons and inlet manifold are of aluminium alloys. Radiators in many cars are of aluminium alloyed with about 1·5 per cent manganese. Bodywork trim can be chosen from anodized aluminium, aluminium-coated plastic, chromium-plated steel or stainless steel. Door handles may be of zinc alloy, plated with nickel and chromium. Radiator grills in some cars are of steel, plated with nickel and chromium, with plastic mouldings for the vertical vanes. Sparking plugs contain an alloy of nickel with manganese, chromium or silicon. The wheels may be of steel, aluminium or magnesium alloys and the wheel balance weights are of a lead–antimony alloy. There is a considerable amount of lead alloy in the battery. In some cars, and especially in Volkswagens, magnesium alloys are used for engine components.

Table 4 shows the weights of materials in the Austin Metro. The different steels are shown separately. Alloying elements in steels, such as chromium, vanadium and molybdenum, are included in the weights of ferrous metals; silicon in aluminium alloys is included in the weight of those alloys. Several other metals, such as tin alloyed with lead in solder, and tungsten in electrical contacts, are included in 'Other metals'.

Most of the metallic materials in motor-cars are alloys. The only pure metal is copper, for electrical wiring, petrol pipes and some washers and gaskets. Although pure metals possess useful properties, such as high

Table 4. Materials used in the Austin Metro

Material	Weight (kgs)	Approximate percentage
Mild steel, for bodywork, tubes, etc.	348·0	49·0
Medium carbon steels	26·0	3·7
Heat-treated alloy steels	42·0	5·9
Spring steel	3·2	0·5
Roller bearing steels	2·7	0·4
Steel sinters	0·5	0·1
Cast iron	110·0	15·5
Aluminium alloys	20·5	2·9
Copper and copper alloys	4·5	0·6
Zinc alloys	1·5	0·2
Lead alloys	9·2	1·3
Other metals	1·0	0·1
Rubber	35·0	4·9
Plastics	35·0	4·9
Glass	30·0	4·2
Other non-metallic materials	41·0	5·8
Total weight	710·1	100·0

conductivity of heat and electricity, they are not often used in structural or mechanical engineering, because their strength is insufficient for the arduous duties which materials are expected to perform. The most important way in which the strength of metals can be increased is by alloying. This in turn opens up possibilities of further improvement of properties by the use of heat treatment, in which the alloy is heated to a selected temperature, below its melting point, and is then cooled at some predetermined rate to obtain specially required properties.

An alloy is an intimate blend of one metal, known as the 'parent' or 'base' metal, with other metals or non-metals. For example, one type of brass contains two thirds copper and one third zinc; steel is iron alloyed with carbon.

Several thousand alloy specifications are in commercial use, kept under review by the British Standards Institution, the International Standards Organization and similar bodies. The documents outline the minimum and maximum limits of composition, and the mechanical and physical properties of each alloy. Specifications relating to manufactured components state dimensional tolerances, surface finish and heat treatment.

HOW ALLOYS ARE MADE

In medieval times, copper alloys were often produced direct, by the smelting of mixed ores of copper and tin, or of copper and zinc. Even nowadays, some alloys are made direct by smelting; thus the complex alloy cast iron is the product of the iron blast furnace; 'Monel', a corrosion-resisting alloy of nickel and copper, used to be made by smelting a nickel–copper–iron ore mined near Sudbury, in Canada. However, most alloys are prepared by mixing metals in the molten state; they dissolve in each other and the alloy so formed is poured into ingot moulds and allowed to solidify, or it is transported molten to foundries where it will be made into castings. Generally the major ingredient is melted first, then the others are added to it. For instance, a plumber making solder would melt lead, add tin, stir and cast the alloy into sticks or bars. Some pairs of metals do not dissolve in this way; for example, if molten aluminium and lead are put together they behave like oil and water; when cast, these metals separate into two layers.

One difficulty in making alloys is that metals have different melting points. Thus copper melts at 1083°C, while zinc melts at 419°C and boils at 907°C; so, in making brass, if we just put pieces of copper and zinc in a crucible and heated them above 1083°C, both metals would certainly melt, but at that high temperature the liquid zinc would boil and the vapour would oxidize in the air. To overcome this difficulty the metal with the higher melting point, namely the copper, is heated first. When this is molten, the solid zinc is added and is quickly dissolved in the liquid copper before very much zinc has vaporized. Even so, allowance has to be made for unavoidable zinc loss – about one part in twenty of the zinc.

Sometimes the making of alloys is complicated because the metal with the higher melting point is in the smaller proportion; for example, one light alloy contains 95½ per cent aluminium (melting point 659°C) with 4½ per cent copper (melting point 1083°C). If the small amount of copper were melted first it would have to be so much overheated to persuade twenty times its weight of aluminium to dissolve in it that gases would be absorbed, leading to unsoundness. In this, as in many other cases, the alloying is done in two stages. First an intermediate 'hardener alloy' is made, containing 50 per cent copper and 50 per cent aluminium, which alloy has a melting point considerably lower than that of copper and, in fact, below that of aluminium. Then the aluminium is melted and the correct amount of the hardener alloy added; thus, to make 100 kg of the aluminium–copper alloy we should require 91 kg of aluminium to be melted first and 9 kg of hardener alloy to be added to it.

In the making of tool steels, cutlery steels, and other alloy steels, manganese, chromium, or silicon are added as ferro-manganese, ferro-chrome, or ferro-silicon, which are cheaper to produce than the pure elements. ('Ferro' indicates the presence of iron; ferro-chromes contain 20 to 55 per cent iron and 80 to 45 per cent chromium.) When alloys are being made a specified working margin is allowed in the composition of the major constituents and the minor impurities. These specifications are always determined so that the alloy will have the required properties and each alloy system must be considered on its own merits. For example in aluminium casting alloys, iron contents of from 0·5 to 1·3 per cent are allowed, as will be seen from the table on page 158. On the other hand the iron content of zinc diecasting alloys is limited to less than 0·1 per cent.

THE MELTING POINT OF ALLOYS

In winter time, when ice forms on the road, salt is thrown down to 'melt' the ice. This practical use of a natural phenomenon demonstrates that a mixture of salt and water has a lower freezing point than that of pure water. The temperature at which pure water just freezes is not low enough to cause the mixture of water and salt to freeze; therefore the salted ice melts. The same effect is to be found in metallurgy, for when one metal is alloyed with another, the melting point is always affected.

The melting point of some alloys can be worked out approximately by arithmetic. For instance, if copper (melting point 1083°C) is alloyed with nickel (melting point 1455°C) a fifty-fifty alloy will melt at about halfway between the two temperatures. However, such an alloy does not melt or freeze* at one fixed and definite temperature, but progressively solidifies over a range of temperature. Thus, if a fifty-fifty copper–nickel alloy is melted and then gradually cooled, it starts freezing at 1312°C; as the temperature falls, more and more of the alloy becomes solid until finally at 1248°C it has solidified completely (*Fig. 10*). This 'freezing range' occurs in most alloys, but it is not found in pure metals, metallic or chemical compounds, nor in some special alloy compositions, known as eutectics, which melt and freeze at fixed temperatures.

The alloying of tin and lead furnishes an example of one of these special cases. Lead melts at 327°C and tin at 232°C. If lead is added to molten tin and the alloy is then cooled, the freezing point of the alloy is found to be lower than the freezing points of both lead and tin (see *Fig. 11*).

* A liquid metal is said to 'freeze' when it becomes solid. Thus copper freezes at 1083°C, just as water freezes at 0°C.

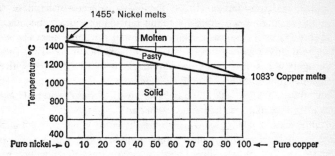

Fig. 10. The melting points of the copper–nickel alloys

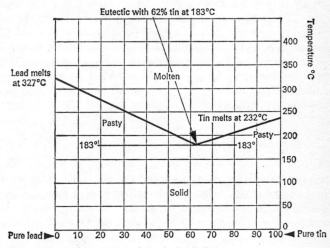

Fig. 11. The melting points of the tin–lead alloys

For instance, when a molten alloy containing 90 per cent tin and 10 per cent lead is cooled, the mixture reaches a temperature of 217°C before it begins to solidify. Then, as the alloy cools further, it gradually changes from a completely fluid condition, through a stage when it is like porridge, until it becomes as thick as paste, and finally, at a temperature as low as 183°C, the whole alloy has become completely solid. By referring to *Fig. 11* it can be seen that, with 80 per cent tin, the alloy starts solidifying at 203°C and finishes only when the temperature has fallen to 183°C (note the recurrence of the 183°C).

At the other end of the series an alloy with 20 per cent tin, a typical solder, starts to freeze at 270°C and is completely solid at the now familiar temperature of 183°C. One particular alloy, containing 62 per cent tin and 38 per cent lead, melts and solidifies entirely at 183°C, without any freezing range. Similar effects occur in many other alloy systems; the special composition in each series which has the lowest freezing point and which entirely freezes at that temperature is known as the eutectic alloy. The word is derived from the Greek *eutektikos*, signifying 'capable of being easily melted'.

By a careful choice of constituents, it is possible to make alloys with unusually low melting points. Such fusible alloys are complex eutectics of four or five metals, mixed so that the melting point is depressed until the lowest melting point possible from any combination of the selected metals is obtained. One fusible alloy contains four parts bismuth, two parts lead, one part tin and one part cadmium. Its melting point is about 70°C, which is less than the boiling point of water. Practical jokers can amuse themselves by casting this alloy into the shape of a teaspoon, which will melt when used to stir a cup of tea. Low-melting-point alloys have been used for anti-fire sprinklers. Each jet contained a piece of fusible alloy, which melted and released water if a fire occurred and the temperature of the room rose.

THE STRENGTH OF ALLOYS

If sugar is mixed with sand, the mixture displays the properties of both ingredients; it is sweet but gritty and the more sand in the sugar the less pleasant is the taste. Metals rarely behave in such a predictable manner. When one metal is added to another, the alloy often has a new individuality which one would not expect from the properties of the two metals.

The laws of heredity teach us that unhealthy parents are likely to beget weakly children, but a weak or soft metal alloyed with another element may produce a strong alloy with strikingly different properties from those of the parent metals. For example, pure iron is a soft metal and carbon in its commonest form is mechanically weak, but as little as half of 1 per cent carbon in iron produces a strong steel which responds to hardening and tempering treatments, and which can be used for making railway lines or rifles.

Again, copper and aluminium are both fairly weak. But the addition of 5 per cent aluminium produces an alloy twice as strong as copper; with a 10 per cent aluminium content, the 'aluminium bronze' produced is three times as strong. Naturally, the temptation would be to go on adding more and more aluminium in the hope of getting still more spectacular increases of strength. Unfortunately, this does not occur. With 10 per cent of aluminium, the aluminium bronze is as strong as mild steel, very tough, and in many ways a valuable engineering alloy. At 16 per cent aluminium, the alloy is about as brittle as a carrot and much less useful. There is a no-man's land between 16 and 88 per cent of aluminium where the alloys are valueless for engineering purposes. (The fifty-fifty hardener alloy mentioned on page 41 is certainly used for making up alloys where its relatively low melting point and its brittleness are convenient. It is, however, only an intermediate product, not an engineering material.) At the other end of this copper–aluminium series, the alloys containing up to 12 per cent copper are stronger than pure aluminium and they, too, are used in industry.

It is rare for all ratios of two metals to have useful properties, though there are exceptions. For instance, the copper–nickel alloys are valuable throughout practically the whole range of composition.

By blending suitable metals it is possible to produce alloys whose strength is much greater than the constituent metals. Furthermore, as will be shown later, many alloys respond to heat treatment processes whereby still better mechanical properties are obtained. Whenever some special property is required – great strength, toughness, resistance to wear, high electrical resistance, magnetic properties or corrosion resistance – it is usually possible to produce it in a carefully chosen, skilfully made and intelligently manipulated alloy.

6

METALS UNDER THE MICROSCOPE

The microscope is the most useful scientific instrument that assists the metallurgist in his everyday work; with its aid a skilled observer can learn a great deal about the structure, the manufacturing history, the effect of heat treatment, and the causes of breakage of any piece of metal or alloy. The microscope in a works laboratory is generally capable of magnifying up to 1000 times, and by the use of different objective lenses a metal can be examined under several successive magnifications. Usually a camera is provided so that the structure of a metal can be photographed as well as observed visually.

As metals are practically opaque, the metallurgical microscope is designed to examine their surfaces by reflected light. A specimen of convenient size, about 10 mm³, is cut off. One face is filed flat and then ground with successively finer grades of carborundum papers which are lubricated with paraffin or water and are supported on plain glass surfaces. Final polishing for optical viewing is continued on cloth impregnated with a slurry of fine alumina or magnesia, to produce a mirror-like finish on the metal. Recently several refinements have been introduced in the preparation of specimens. Silicon carbide, boron nitride and diamond dust are being used as the abrasives, which helps to give better reflective surfaces than before. The cloths used for polishing are now selected from a range of textile materials, chosen to suit the hardness of the specimen.

First the polished surface is examined under the microscope to see if there are any cracks, inclusions or holes. *Fig. 12a* shows the appearance of a piece of polished wrought iron such as was made in the nineteenth century for chains, horseshoes or the gates of ancestral homes. The dark streaks are inclusions of slag which was always present in this metal due to its method of manufacture, described on page 112. Copper of certain grades may contain small particles of oxide which can be identified microscopically. If a metal has failed in service, the position and shape of the crack can be examined and the cause of the failure may be diagnosed; for example, it may be found that the crack began at a piece of slag or other inclusion.

Such an examination, useful though it is, does not indicate the structure or composition of an alloy, and it tells little of the success or otherwise of heat treatment. Much information can, however, be obtained by etching the polished surface of the metal, and re-examining it under the microscope. The specimen is immersed in a chemical solution, chosen according to the metal that is to be examined, and the particular constituent it is desired to observe.

For general examination, steel is usually etched in a mixture of 98 parts alcohol and 2 parts nitric acid, though if one particular constituent, known as 'cementite', has to be identified, the steel is etched in boiling sodium picrate solution which makes the cementite appear black. The time of etching varies from a few seconds to half an hour, depending on the alloy and on the solution employed. Sometimes more sophisticated methods are used, such as electrolytic attack or distilling away some of the metallic surface in a vacuum or by slightly heating the specimen to form thin films of oxide.

THE GRAIN STRUCTURE OF METALS

Microscopic examination of the etched surface of a polished metal reveals that it is built up of innumerable small 'grains'. *Fig. 12b* shows the same piece of wrought iron as in *Fig. 12a*, but this time the metal has been etched in a dilute solution of nitric acid in alcohol. It will be understood that the illustration represents a section cut through the metal and that in reality the grains are three-dimensional.

Like many other technical words 'grain' has a somewhat different meaning from the popular interpretation, which connects grains with sand, salt, or sugar. A grain of metal differs from one of sugar in at least two respects. Firstly, each grain of sugar is brittle and can be crushed into powder, whereas metal grains are usually ductile. Secondly, little cohesion exists between the grains of a lump of sugar, while in a block of metal great force is needed to separate the grains. Sometimes this cohesion of the boundaries may be weakened; for example, when one part in ten thousand of bismuth is present in copper or gold, the bismuth distributes itself at the grain boundaries, thus reducing cohesion and causing the piece of copper or gold to be brittle.

On etching the metal, the grain boundaries are more readily attacked

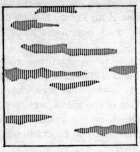 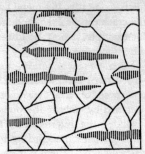

a Before etching   *b* After etching

Fig. 12. Wrought iron

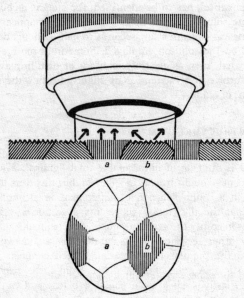

Fig. 13

than the interior of the grains. Tiny channels are eaten away at the extremities of each grain; when the etched metal is examined under the microscope, light falling on these channels is scattered, so that the boundaries around the grains appear dark. Under some conditions of etching, whole grains appear contrasted in tone, which may be even more distinguishing than the demarcation of the boundaries. *Fig. 13*

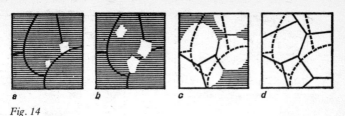

Fig. 14

illustrates the cause of this contrast. It will be seen that the grains have been attacked by the etching solution in different ways. When light falls on the grain marked '*a*' most of it will be reflected back through the objective of the microscope and consequently this grain appears bright. On the other hand light falling on grain '*b*' is reflected sideways and so the whole of this grain appears darker.

Although the grain structure of most metals can be revealed only after etching and with the aid of the microscope, there are a few examples where large metal grains can be seen with the unaided eye. When brass articles such as door knobs, or the long brass handles at the entrances to public buildings, have been in use for a year or so, the constant rubbing with human hands polishes and etches the brass, so that the grains are revealed as small patches or spangles of slightly varying colour tones. In such brass knobs and handles the grain size is comparatively large, of the order of two millimetres. Large grains of zinc can also be seen on galvanized steel articles.

However, in most metals and alloys the grain structure can be distinguished only by the use of a microscope. For example, some steels, in the condition in which they are used in industry, have grains about one fourth of a millimetre across. The grain size of a metal depends on the casting temperature, the impurities present, and the mechanical working and heat treatment to which the metal has been subjected; this is one of the important ways in which metallurgists can fit metals for the tasks they will have to perform in service. In general, a metal with fine grains will be somewhat harder and stronger than one with coarse grains.

After a metal has been 'cold worked' – drawn into wire or rolled into sheet – it is possible to bring about the birth of new grains by heating the metal. This is illustrated in *Fig. 14*, and the process is known as 'recrystallization'. The first effect of heating is to form small, new grains as shown in white in *a*, and these rapidly enlarge until further growth is restricted by one new grain meeting another as shown in *b* and *c*. Ultimately the original system of grains is obliterated and the new,

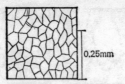

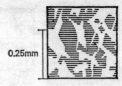

Fig. 15. Copper–nickel alloy   *Fig. 16*. 60/40 brass

recrystallized structure is shown in *d*, the original grains being indicated in the drawing by dotted lines.

On continuing the heating, adjustments of the boundaries may take place, resulting in the further growth of some grains at the expense of others. The eventual size of the grains in a piece of metal depends on the amount of deformation previously existing and on the time and temperature of the heating process. By special means, a piece of metal can be treated so that it becomes one large grain, known as a single crystal.

The production and study of single crystals has provided much valuable information to scientists. Some single crystals, diamonds for example, are produced by nature; single crystals of copper sulphate can be made from a solution in any laboratory. Large metallic single crystals, often possessing remarkable properties of ductility in one or more directions, can be 'grown' from molten metal by using a small crystal of the metal to provide a 'seed' about which the molten metal solidifies. Nowadays large single crystals are being manufactured; those from which silicon chips are cut are sometimes two metres long and 20 centimetres in diameter.

THE STRUCTURE OF ALLOYS: SOLID SOLUTIONS

The properties of any metal are altered when it is alloyed with another, and this makes one wonder whether microscopic examination will reveal structural differences between pure metals and alloys. In some cases such differences can be observed; thus, two constituents can usually be identified in the lead–tin or the iron–carbon alloys. But other alloys may consist of the polyhedral grains characteristic of a pure metal, and only one constituent is discernible. For example, the microstructure of a copper–nickel alloy is as shown in *Fig. 15*, which may be compared with that of wrought iron in *Fig. 12b*.

If this alloy were composed of grains of copper mixed among grains of nickel, it would be possible to distinguish them by colour alone. Clearly

Table 5. The solid solubility of various metals in magnesium at room temperature and at 300°C

| Element | Solid solubility in magnesium (Weight per cent) | |
	Room temperature	300°C
Aluminium	2·3	5·3
Calcium	about 0·1	0·18
Copper	under 0·1	0·1
Lead	3·7	16·0
Manganese	nil	0·1
Silver	1·5	3·6
Zinc	1·7	6·0

then, something must have happened to the copper and nickel atoms to mingle them so closely that microscopic examination cannot reveal the individual metals.

The two metals are said to exist in a state of 'solid solution'. It may seem strange that one solid metal can exist in solution in another, but there is a wider definition to the word 'solution' than merely 'something dissolved in a liquid'. A solution may be described as an intermingling of one substance in another so closely that the dissolved substance cannot be distinguished or separated by mechanical means. This description can be applied to the condition of the solid copper–nickel alloy just as it is applied to sugar dissolved in water.

Only a few pairs of metals, such as copper and nickel, can exist in solid solution throughout the whole range of possible compositions, but most metals can contain at least some of another metal in solid solution. In an alloy where the metals do not show complete solid solubility, separate constituents may be recognized by examining the etched alloy under the microscope. *Fig. 16* shows the microstructure of a brass containing about 40 per cent zinc, similar to that used for domestic water taps or brass nuts. The two constituents are two solid solutions, each of different composition.

Just as tea dissolves more sugar when hot than when cold, so a metal can usually retain more of another metal in solid solution when it is hot than when cold. *Table 5* shows the solid solubility of various metals in magnesium at room temperature and at 300°C. It will be seen that the solid solubility of each metal is greater at 300°C.

INTERMETALLIC COMPOUNDS

In the aluminium–copper alloys, solid aluminium at about 530°C can retain 5 per cent of copper in solution, while at room temperature it can normally hold less than half of 1 per cent. Therefore, if an alloy containing, say, 4 per cent copper is slowly cooled from about 530°C, it comes to a stage, at just below 500°C, when it can no longer hold as much as 4 per cent copper and, as the temperature falls further, less and less copper can be retained in solution. The surplus copper does not separate as distinct grains of that metal, but in the form of an 'intermetallic compound' to which the symbol $CuAl_2$ is given.*

As the temperature becomes progressively lower, more and more copper comes out of solution and forms $CuAl_2$, so that finally the slowly cooled alloy at room temperature consists of a background of aluminium containing only a small amount of copper in solid solution, together with a number of particles of the intermetallic compound $CuAl_2$ dispersed throughout the alloy. When isolated, this compound is found to have characteristic properties; it is, for example, extremely hard.

Plate 23a shows the microstructure of a lead–tin–antimony alloy, such as is used for bearings. The 'cubes' are of an intermetallic compound of tin and antimony. When the composition of an alloy is such that its structure consists of a matrix of solid solution and particles or grains of an intermetallic compound, the alloy is likely to be useful, for it combines the toughness of the solid solution with the hardness of the intermetallic compound. But if the intermetallic compound predominates, that alloy displays hard and brittle properties and consequently may be unserviceable.

EQUILIBRIUM CONDITIONS

The behaviour of the aluminium–copper alloys will be referred to again on page 161 when age-hardening is discussed, but it may be remarked here that the complete separation of constituents is effected only on slow cooling. If the aluminium alloy containing 4 per cent copper is quenched in water from 500°C so that it is rapidly cooled to room temperature, the copper does not have time to come out of solid solution and the aluminium at room temperature is forced to hold a surplus of copper in solid solution. In other words it is 'super-saturated' and it may be some days afterwards before the copper atoms spontaneously separate. In

---

* $CuAl_2$: its composition by weight is 54 per cent copper, 46 per cent aluminium. The symbol denotes that one copper atom is intimately associated with the two aluminium ones.

some other alloys treated in this way a state can be reached by quenching where adjustment can be attained only by warming the alloy, which gives opportunity for the separation to take place.

Many of the phenomena of metallurgy may be attributed to the sluggishness of alloys in attaining equilibrium. The hardening and subsequent tempering of steel and the age-hardening of aluminium alloys depend on the fact that rapid quenching makes their condition different from that produced by leisurely cooling in a furnace.

THE STRUCTURE OF EUTECTICS

There is another type of constituent seen under the microscope which may appear in those series of alloys which form eutectics. It will be remembered from page 43 that in certain alloys there exists one particular composition which melts at a lower temperature than any other alloy in that series. This alloy is known as the eutectic. The structure of eutectics may consist of alternate thin layers of the metals concerned, or in other cases small globules of one metal embedded in a matrix of the other. For example, the silver–copper eutectic is composed of 72 per cent silver and 28 per cent copper. If an alloy in such a series is not of eutectic composition, the structure as seen under the microscope consists of one of the metals and some eutectic. Thus an alloy of 10 per cent silver and 90 per cent copper has a structure of grains of copper plus a small amount of eutectic. Similarly at the other end of the series, an alloy with 90 per cent silver and 10 per cent copper consists of eutectic plus silver. The nearer the composition approaches that of the eutectic, the greater the proportion of eutectic seen when the alloy is viewed under the microscope.

THE BIRTH OF GRAINS FROM MOLTEN METALS

Most people have seen the attractive patterns formed in winter when water vapour freezes on windows. A similar beautiful structure may be formed when a liquid alloy solidifies, though the alloy has to be polished, etched, and examined under the microscope to make its structure apparent. When a molten alloy begins to freeze, minute crystals form at various points in the liquid and these start to grow by developing branches in certain directions, as shown in *Figs. 17a* and *b*. These tree-like formations are known as 'dendrites'. When the arm of a dendrite meets that of another, as in *c*, outward growth is restricted, but the spaces between the branches continue to fill in until all the metal is solid, as

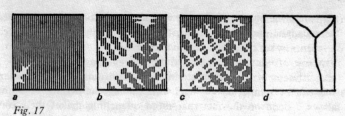

Fig. 17

in *d*. Photographs of dendrites, in antimony, and in a copper–silver alloy, are shown in *Plates 22a* and *23b*.

When a solid piece of *pure metal* is sectioned, polished, etched, and examined under the microscope, no sign of the dendrites can be seen, because the metal is uniform. But when a cast *alloy* is solidified and examined microscopically, evidence of the dendrites can usually be seen, for the composition of the first part of the alloy to freeze differs from that which finally freezes, and what is known as a 'cored structure' is produced. If such an alloy is then worked either hot or cold and is heat treated, this cored structure may be gradually eliminated and a homogeneous solid solution may be formed, having the uniform structure which has been shown in *Fig. 15*.

The techniques of metallography have made many advances. Fifty years ago a microscope that could give a magnification of 1500 was considered powerful; then it became possible to photograph the microstructures of metals at magnifications of 7000 by the use of ultra-violet light. Next, microscopes using beams of electrons were developed, and soon magnifications of over 100 000 became possible. The photograph of dislocations in cold-worked aluminium on *Plate 24a* was taken at a magnification of 50 000.

The latest electron microscopes can magnify more than a million times and can resolve details in very thin specimens down to less than two ten-millionths of a millimetre, where individual atoms can be viewed. *Plate 4* shows a Japanese microscope in a Californian laboratory. The tank at the top left of the picture is the voltage generator which provides the 400 000 volts, to accelerate the electron beam. The column above the operator contains the series of electromagnetic 'lenses' that focus the electron source on to the sample and then magnify the image on to a fluorescent screen. There is a high vacuum in the column because the electrons can only travel under those conditions.

Such a microscope has yielded a great deal of understanding of phenomena in metals, such as crack propagation, defects in crystal structures, the effects of impurities and of nuclear radiation on metals. Preparing the samples of the materials to be examined is a skilled task, sometimes taking several days, because the region examined is only about one ten-thousandth of a millimetre thick.

A different type of instrument, the scanning electron microscope, examines surfaces in great detail and it has the advantage that it does not require the fantastically thin specimens that must be produced for the microscope described above. The SEM provides resolutions up to a million times and has a large depth of focus to 'see' inside deep cracks. Many research activities have benefited from the use of the SEM in studies of crystal growth, corrosion, electroplated deposits and bonding processes.

The electron probe analyser reveals the constituent elements on and near the surface of materials from the characteristic X-rays that are emitted when an energetic electron beam is scanned over the area. By this means chemical 'maps' are plotted, showing the distribution of elements in micro-areas less than a thousandth of a millimetre across.

In recent years the precise study of the surfaces of materials has become an important branch of science. Another device which has contributed to these studies is the Scanning Auger microscope, illustrated in *Plate 5*. A beam of high-energy electrons is fired at the material to be examined, causing low-energy electrons to be ejected from the first few atomic layers. The energy of these electrons is very specific to the atoms from which they originated and thus, so to speak, every element has its own 'Auger trademark'. At first the Auger instruments were used to monitor the cleanliness of semi-conductors and of surfaces prior to the deposition of coatings in individual processes, such as the galvanizing of steel. More recently, however, Scanning Auger microscopes have proved invaluable in the search for new superconductors, described on page 289. By the use of such methods, scientists are able to experiment on the scale where most of the basic phenomena occur.

7

THE INNER STRUCTURE OF METALS

All chemical elements, including the metals, are composed of atoms, which are the smallest particles retaining the individual characteristics of the element in question. The atoms are so small* that they cannot be seen, except by the most powerful electron microscopes. Using such instruments, physicists have been able to widen our knowledge about atoms and this has helped to explain the behaviour of metals. It is important to know how the atoms are arranged in a grain of metal. Are they all piled at random, or do they exist in a regular and orderly formation? Are the atoms of all metals exactly the same size, or does an atom of, say, lead occupy less space than an atom of magnesium?

Even in the later years of the nineteenth century, scientists had evidence that the atoms in metals were arranged in a regular geometrical fashion, that is, metals belonged to the class of substances which are crystalline. But it was not until 1911 that Max von Laue produced proof of this belief by the examination of metals with X-rays.

The conception that metallic grains are crystals is often a stumbling-block to those who are beginning to study the science of metals. Most people associate crystals with such substances as quartz, diamonds, or copper sulphate, which are hard and sparkling, but this does not imply that all crystalline bodies have these characteristics. The basic definition of crystalline nature does not concern the outward appearance of a substance, but arises from its inner symmetry. There are many quite soft substances which are truly crystalline, whilst some hard sparkling sub-stances, such as glass, are not true crystals.

Metallurgists and physicists have discovered the different 'lattice' structures which the atoms adopt in the various metals, the distance

---

* It has been calculated that a cubic millimetre of copper contains about 84 693 000 000 000 000 000 atoms, that is about 15 thousand million atoms for every person in the world.

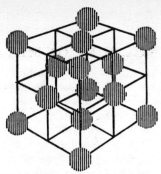

Fig. 18. Face-centred cubic lattice

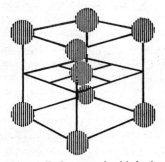

Fig. 19. Body-centred cubic lattice

between neighbouring atoms, and the space which the individual atoms occupy. In aluminium, copper, nickel, lead, silver, gold, platinum, and several other metals, the atoms are spaced evenly in rows at right-angles to each other. This can be likened to atoms arranged at each corner of millions of adjoining cubes, while other atoms occupy positions at the centre of each of the cube faces. This particular atomic lattice pattern is known as 'face-centred cubic', and is illustrated in *Fig. 18*.

In iron, when at room temperature, and in several other metals such as vanadium, tungsten, molybdenum, and sodium, the atoms are disposed in another pattern. There is again an atom at each corner of each imaginary cube, but instead of other atoms occurring at the middle of the cube faces, a single atom is located at the centre of every cube. This structure is known as 'body-centred cubic' and is illustrated in *Fig. 19*. Iron is particularly interesting, for at room temperature its atoms are arranged in the body-centred cubic form, but at 906°C the atoms reshuffle into the face-centred cubic pattern, while at a still higher temperature of about 1400°C the iron atoms change back to a body-centred cubic lattice.

In zinc, magnesium, titanium and cadmium the atoms are arranged in a hexagonal pattern, while some other metals have still more complex lattices. Most common metals have either face-centred cubic, body-centred cubic or hexagonal atomic lattices.

After having discovered how the atoms of pure metals were arranged, investigators turned their attention to the lattice patterns of alloys. From microscopic examination they already knew that the grain structures of alloys were different from that of the metals of which they were composed, and so it seemed likely that some parallel difference might be found when the atomic lattices of alloys were investigated. There were interesting problems which required elucidation; for example, what happens to the atomic arrangement when a metal which crystallizes in the face-centred cubic pattern is alloyed with one of a hexagonal type? Can any light be thrown on the cause of the increase of strength which occurs when one metal is alloyed with another? In the following discussion we consider the behaviour of one alloy, brass.

THE ATOMIC LATTICE PATTERNS OF BRASS

The atoms in pure zinc are disposed in a hexagonal arrangement whereas copper atoms form a face-centred cubic lattice structure. The space occupied by the zinc and copper atoms differs, that of zinc being about 13 per cent larger than that of the copper atoms. When only small amounts of the zinc are present in brass, the prevailing atomic arrangement is similar to that of copper, which means that zinc atoms, in solid solution, have to adapt themselves to the face-centred cubic pattern. Each zinc atom takes the place of one copper atom, but because the space occupied by the zinc atom is greater than that of copper, the face-centred cubic lattice is distorted at the point where the 'stranger atom' is introduced, and the whole of the lattice becomes slightly larger. Perhaps a comparison of *Fig. 20* with *Fig. 18* will demonstrate this better than words.

When progressively increasing amounts of zinc are alloyed with copper, the brass becomes increasingly hard; this can be explained by referring to the drawing of the stranger atom. The introduction of a new atom of different size from the rest brings about a condition of distortion in the lattice, and this leads to a greater resistance to deformation than occurs in the pure metal. The strengthening by alloying can occur

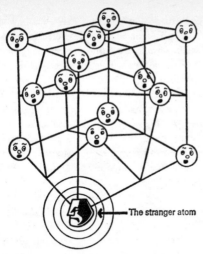

The stranger atom

Fig. 20

whether the stranger atoms are larger or smaller than those of the original metal, for distortion of the lattice occurs in either case.

As more zinc is added, the face-centred cubic arrangement of the atoms becomes increasingly distorted till, when about 36 per cent of zinc is present, the lattice becomes unstable and here and there another form comes into existence, which is body-centred cubic. Up to this point the strength and hardness of the brass gradually increase with rising zinc content, but at the 36 per cent composition the properties are sharply altered and further additions of zinc bring about a more rapid rise of hardness than before.

A change can also be noticed when the alloy is examined under the microscope. Up to 36 per cent zinc, the brass is in the form of a solid solution in which no direct indication of the presence of the zinc can be observed except by some change of colour. With over 36 per cent of zinc a new constituent or 'phase' begins to appear and can be seen under the microscope (see *Fig. 16*). It is necessary to distinguish between these phases, and they are given Greek letters to designate them. According to the usual practice, the first phase is called alpha (α), the next beta (β); a brass containing 36 to 42 per cent of zinc includes both alpha and beta phases, and is called an alpha-beta brass. When still more zinc is added, further phases appear with different atomic lattices; and with each new structure the mechanical properties of the brass alter sharply. In all, there are five different constituents which can exist, though not more than two at once, in the copper–zinc alloys. The effects of increasing amounts of zinc alloyed with copper are summarized in *Table 6*, which compares the composition of the alloys with their microstructure, atomic lattices, and mechanical properties.

The copper–zinc alloys are rather complicated because so many atomic reshufflings occur. Some alloy systems, such as the copper–tin series, are even more complex; others, such as copper–nickel alloys, are straight-forward because both metals have face-centred cubic atomic lattices, their atoms are of similar size, and they are able to form a continuous range of solid solutions, the lattice dimensions changing gradually, from pure copper to pure nickel.

Five points concerning the atomic structure of metals are summarized below:
1 The atoms of each metal have a characteristic size, differing from all the others.
2 The atoms in each grain of a solid metal are arranged in a regular pattern. There are several types of patterns, the three most important for metals being known as face-centred cubic, body-centred cubic, and hexagonal.
3 When one metal exists in solid solution in another, the atoms of the added element usually take their place on the atomic lattice of the parent metal, in spite of the fact that the atomic size of the added metal is different and its normal atomic lattice pattern may be different. Such intrusion causes the lattice to be distorted and the alloy is harder and stronger than the parent metal.
4 When an alloy is made from two metals which possess a markedly different size of atom and which crystallize in different atomic patterns, the range of alloys divides into a number of phases. The occurrence of a new phase is characterized by a change in the atomic arrangement and is also evident when the alloy is examined under the microscope.
5 When an intermetallic compound is formed, the various kinds of atom present build an entirely new type of lattice, which is sometimes quite complicated and is different from that of either ingredient.

IMPERFECTIONS IN CRYSTALS

From the previous description of the crystalline form of metals, it might be concluded that such crystals are perfectly regular arrangements of

Table 6. Some changes produced by alloying zinc with copper

	Atomic structure				Mechanical properties *			
Composition of alloy	Description of lattice	Spacing of atoms Ångström Units	Crystal structure as seen under microscope	Zinc per cent	Diamond pyramid hardness number	Tensile strength — Tons per sq. inch	Tensile strength — Newtons per sq. mm	Ductility (expressed as per cent elongation)
Pure copper	Face-centred cubic	3·615	Grains of pure copper	—	53	15	230	45
Copper alloyed with up to 36% zinc	Face-centred cubic pattern which progressively increases in size	3·615 with 0% zinc, increasing to 3·698 with 36% zinc	Grains of solid solution of zinc in copper (known as alpha phase)	10 20 30	60 62 65	18 20 21	280 310 325	55 65 70
Copper with about 36% zinc	A new structure appears (body-centred cubic) in addition to the original face-centred cubic structure	3·698 for alpha constituent / 2·935 for new (beta) constituent	Small quantities of a new constituent appear (known as beta phase)	36	70	22	340	60
Copper with 36 to 42% zinc	With increasing zinc more of the alloy consists of the new constituent having body-centred cubic lattice	On change to body-centred cubic lattice the atoms are 2·935 units apart	The beta constituent increases in quantity while the alpha constituent diminishes	40	85	27	395	45
Copper with 42 to 52% zinc	The lattice is entirely body-centred cubic	2·935 with 42% zinc, increasing to 2·941 with 52% zinc	The structure is entirely beta	45	90	28	410	20

* Units of hardness, strength and ductility are explained on pages 92–9; Ångström units on page 292.

atoms, like marbles in a game of Solitaire. However, a great deal of evidence over the past forty years has confirmed that metal grains or crystals are far from perfect and that many of their attractive mechanical properties are due to imperfections. In particular it leads to an explanation of why most metals are tough and not brittle, why they will stand heavy loads, extend a little and then stop extending; it also helps us explain why metals will endure shock loads and reversals of stress better than most non-metallic materials, which do not have these kinds of imperfections.

If one considers the tree-like formations known as dendrites which are illustrated in *Fig. 17*, page 54, it is easy to imagine that a slight bending of each branch while growing from the molten metal leads to a lack of registry when the branches meet and the metal finally freezes. Because the branches are linear, these misfits tend to be in parallel lines between successive branches. Another type of irregularity is a block arrangement known as a sub-grain, about 0·001 mm across, where each block of atoms is slightly deranged relative to its adjacent blocks, like a jerry-built brick wall.

The late Sir Lawrence Bragg was the first man to demonstrate the kinds of disarrangements which a regular pattern of uniform atoms might undergo at a metal grain boundary. He displayed this in a simple and rather beautiful experiment, using bubbles on the surface of water. This is the so-called 'bubble-raft experiment'. *Plate 6* shows the way in which a large group of bubbles arrange themselves. It might be expected that the configuration of the bubbles would be in absolutely straight lines but the photograph shows a slight change in direction near the bottom right hand corner and a Y-shaped join in the top left hand quarter of the photograph. These illustrate that what can happen with bubbles can happen with atoms, though it must be realized that this is a great simplification of the three-dimensional disarrays which occur in the structure of metals.

The photo-micrographs in *Plates 24a*, *b*, *c* and *d* show some imperfections or dislocations; furthermore they illustrate how the electron microscope has made it possible to examine metals under the magnifications that would have been impossible forty years ago. *Plate 24a* is an electron microscope picture, at a magnification of 50 000. A piece of aluminium has been given 10 per cent reduction in thickness. The fuzzy dislocations, like tangled wool, are disposed in a cellular arrangement around sub-grains, the white areas, where there are very few dislocations. When one remembers that normal working operations on such a metal cause reductions of 25 to 50 per cent, it will be realized that such heavily worked metals would show many more dislocations.

Plate 24b, which looks like part of a painting by Miró, shows dislocation lines that start and finish at black spots, which are in fact particles of aluminium oxide entrapped in a sample of brass. The photograph, at a magnification of 10 000, shows how the oxide particles 'trap' the dislocations. The effect is to harden and strengthen the alloys.

Plate 24c shows an iron alloy containing 3·25 per cent silicon, taken at a magnification of 2200. The three parallel bands are slip-bands, which will be discussed again in Chapter 9. The dotted line that crosses the slip-bands is a disclocation.

Plate 24d is a photo-micrograph of the iron–silicon alloy referred to above. It has been cold rolled to give a 4 per cent reduction in thickness and annealed until recrystallization commences. The dark areas are those where the effect of dislocations remains and the light areas show that, on annealing, new grains have begun to show.

Although these effects are loosely described as dislocations, there are several associated effects, all of which have a profound influence on the strength of metals:

1 Point defects, caused by atoms which are small enough to squeeze into the interstices between larger atoms.
2 'Vacancies', where atoms are missing.
3 Line defects, called dislocations.
4 Surface and inter-face defects, which include grain boundaries and sub-grain or block boundaries.

A knowledge of the inner structure of metals is of more than theoretical interest, for by helping us to understand how metals and alloys are built up, it enables us to exert a more precise and comprehending control over alloy composition and heat-treatment than would otherwise be possible. Also it helps to explain what happens when metals are hardened or stressed. In all branches of science, investigations which at first appeared to be of merely academic interest have often proved to be of great practical benefit, and the study of the inner structure of metals was no exception to this rule.

8

SHAPING METALS

Consider the making of a sewing needle. A white-hot ingot of 0·8 per cent carbon steel is forced between pairs of rotating, grooved rolls. This process reduces the thickness and increases the length of the steel, giving a square-sectioned bar, and is followed by a further rolling treatment, using grooved rolls of such a shape that the bar is made into rods of circular section about 12 mm diameter. In the next stage the metal, when cold, is drawn through successively smaller holes in hard steel dies which reduce it in diameter, so that eventually it becomes wire of the same diameter as the needle. The wire is then finally annealed to soften it, special precautions being taken that none of the carbon of the steel is lost by the action of the furnace atmosphere. The wire is then cut to a length just over twice that of a needle and each end is pointed by grinding, in a continuous process, on a rapidly rotating emery wheel. The middle part of the annealed wire is stamped so that it has the form of two needle heads joined together and an eye is then pierced in each head (*Fig. 21*).

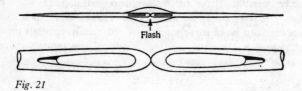

Flash

Fig. 21

The twin needle is broken into two and the 'flash' resulting from the stamping operation is ground away. The pieces of wire now begin to look much like needles, although they are still so soft that they can easily be bent double. In the next operation the steel is hardened by heating the needles to a bright red heat and then quenching. The steel is now extremely hard, and so brittle that a handful of needles can be snapped like macaroni. Before the needle is suitable for use it must be tempered, so that whilst maintaining much of the hardness, the brittleness is removed. The tempering involves heating the steel to a temperature of a little over 200°C.

The next process is remarkable. Thousands of needles are placed on a canvas sheet, covered with emery powder and soft soap, and the canvas is rolled up like a roly-poly pudding and revolved between weighted rollers for many hours. This traditional process is still used but in modern times a great deal of development work has been carried out to replace the roly-poly method with one of the modern bulk finishing operations, thereby reducing the time and manual effort required.

Every metal article has more or less as fascinating a manufacturing history as a sewing needle. Although a multitude of shaping processes are used, they can be grouped into five classes: shaping from molten metal, from hot solid metal and from cold metal. These three are discussed in the present chapter. The joining of metals is described in Chapter 21 and powder metallurgy in Chapter 22.

SHAPING FROM MOLTEN METAL

Remains of an ancient metal-melting furnace with cup-shaped crucibles dating to 3000 B.C. have been found at Abu Matar, near Beersheba. Today the same principle is used for melting and holding up to a few hundred kilograms of molten metal. In the old days the metal was heated by burning charcoal. During the last century coke, gas and oil have been used.

For melting large amounts the metal is contained in a shallow bath of comparatively large area; hot gas or an oil flame plays on the sloping furnace roof and 'reverberates' heat on to the surface of the metal.

In the twentieth century electric arc and electric resistance furnaces began to be used for metal melting. Low-frequency electric induction furnaces have been widely exploited and high-frequency induction furnaces have become standard equipment for melting many high-quality alloys. There have been developments in the casting technology of the newer metals, particularly molybdenum, titanium, zirconium, and some of the 'super-alloy' steels. An important step has been the perfection of consumable-arc-vacuum and protective-atmosphere furnaces. More recently beams of electrons have been used to ensure the highest purity of metal during the melting operation.

MAKING A SAND CASTING

There are several ways of producing castings from molten metal; those made from sand moulds are perhaps the best known. Let us assume that a casting is to be made shaped like *Fig. 22a*, and that only the simplest equipment is to be used. A solid pattern or model of the letter M is first prepared and from it an M-shaped cavity is made in moulding sand. A pair of metal boxes open at the top and bottom will contain the sand and help it hold together. One box is placed on a board and the pattern of the letter M is put face downwards on the board in a central position. The box is filled with sand, which is rammed firmly round the pattern, another board is placed on top of the moulding box, which is then turned over so that the pattern is uppermost (*Fig. 22b*).

The pattern is now carefully extracted from the sand, leaving an M-shaped cavity. It would be possible to make a casting by pouring liquid metal into this open mould, but the top face of such a casting would not be flat, because of shrinkage of the metal as it sets. In practice, moulds are made in two parts, and for casting the letter M the upper half consists of a moulding box, containing sand with a carefully smoothed surface. This is placed on top of the first moulding box containing the impression of the letter M; the cavity is now completely enclosed in sand (*Fig. 22c*).

A channel or 'runner' must be cut through the sand in the one half of the mould so that liquid metal can be poured down this runner and flow into the cavity (*Fig. 22d*). One or more channels, called 'risers', must also be made; the cast metal enters the runner, fills the mould and rises up the riser which assists the complete filling of the mould and helps to ensure soundness in the casting by providing an outlet for air in the mould. When the metal has solidified and the two halves of the mould are separated, the casting remains attached to necks of metal, representing the runner and riser which are subsequently cut off (*Fig. 22e*).

The description above has been of a simple 'one-off' casting made under primitive conditions. Anyone who has not seen modern foundries at work may be surprised to know that complicated sand moulds can be made which, when treated carefully, can be inverted and will stand up to the stream of molten metal without being washed away. The choice of the right kind of sand and the strengthening of it with bonding materials make this possible. Apart from these considerations, the sand must be packed or rammed to the correct degree. If it is rammed too tightly, air contained in the cavity, and water vapour from the heated sand, are trapped in the molten metal, causing unsoundness. On the other hand,

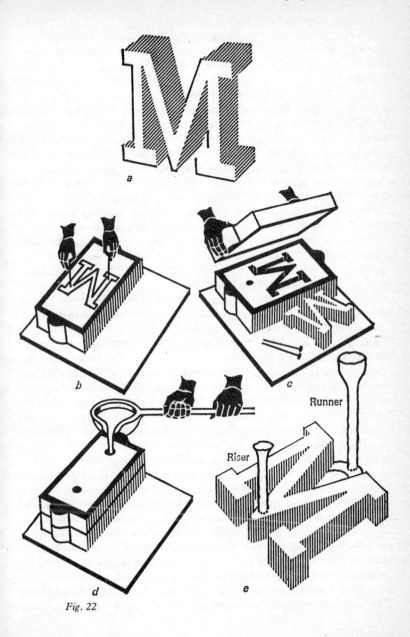

Fig. 22

insufficient ramming may lead to the mould being washed away by the cast metal. Because of contraction which occurs on solidification, extra metal is added or 'fed' after the main bulk has been poured. In casting marine propellers, described later, the feeding continues for several hours.

When a casting has to include holes or other recessed features, these portions are made by separate cores which are fitted into the two halves of the mould after the pattern has been taken out. Cores are frequently made of sand bonded with oil and baked to give them strength and rigidity, or bonded with chemicals that harden in air. A complicated large mould may have over a hundred cores and take many hours to make.

At one time most moulds were made of sand bonded with the natural clay which it contained when quarried. (As ex-makers of sandcastles will remember, sand from the sea-shore does not bind well together because its natural clay bond has been washed out by the salt water.) There are extensive deposits of silica sands very suitable for moulding. The best known are found in the Mansfield, Leighton Buzzard and Bromsgrove areas, which give their names to the sand, although the quarries extend over a much wider area of the countryside.

With the advent of automatic moulding machines more precise control of sand properties was necessary and the introduction of synthetic sands became common. However, many new chemical materials have been introduced to strengthen the sands, including 'shell moulding' resins which harden when heated, and materials in the so-called Hot Box process, where the moulds are cured in contact with heated metal patterns or boxes. More recently 'cold curing' resins have been developed; by correct selection of materials and catalysts it is possible to obtain mixtures which either harden in a matter of minutes or, using special dispensing equipment, they may be hardened in seconds after the sand has been applied to the pattern.

For the production of big quantities of similarly shaped castings, foundries are mechanized, applying the principles of mass-production. The patterns are of metal and the moulding boxes are mounted on a conveyor. The sand, having been mechanically mixed and reconditioned, is automatically flung into the moulding boxes which are vibrated on machines so that the sand is consolidated correctly. The moulds are assembled, and while they are still moving along the conveyor belt the metal is poured into them; after the belt has moved forward and the casting solidified, the mould is automatically tilted and the solid casting removed. The mould box continues and the sand falls out to be reconditioned.

Components such as racing engine cylinder heads are subjected to unusually severe working conditions and the need for the utmost reliability is paramount. Among the world-wide efforts to attain perfection, the Cosworth casting process, developed in Britain, involves the use of high purity alloys without the addition of any scrap, and a special design of the moulds prevents contamination. Zircon sands are used in mould manufacture; these are expensive but are much more refractory and consistent than silica sands. In the Cosworth process, molten metal is withdrawn from the centre of the melt, by an electromagnetic pump, so that turbulence and oxidation are eliminated.

CASTING A LARGE MARINE PROPELLER

Screw propellers were developed during the nineteenth century. At first they were simple shapes with blades of constant pitch from root to tip and they were generally produced in the engine-builders' own foundries. Now propeller shapes have become more sophisticated and, before production, accurately constructed models are studied scientifically in testing tanks before designs are finalized and the propellers cast in specialized foundries.

Early propellers were made in cast iron or cast steel but towards the end of the nineteenth century 'high tensile brass', containing iron, manganese or aluminium, was introduced for its strength and resistance to corrosion. In the 1950s the power and size of vessels increased considerably with the introduction of 'V.L.C.C.s' (Very Large Crude Carriers) and container ships. With these developments came the introduction of high-strength propeller alloys with great resistance to cavitation and corrosion fatigue (pages 176, 234). These are aluminium bronzes with the addition of manganese, or nickel and iron. *Plate 7* shows a propeller fitted to a 'Europa' class V.L.C.C. of 380 000 tonnes dead weight. The cast weight of the propeller was 93 tonnes and the finished weight 70 tonnes.

The material used for moulding is a mixture of silica sand and cement. Pits are necessary for moulding large propellers, partly for reasons of safety and partly to bring the top of the mould to a reasonable height for working. A series of massive concentric slotted cast-iron rings are bolted into the floor of the pit, to enable the mould to be held rigidly.

The mould for each blade of the propeller is made in two halves, the 'bed' and the top. The upper surface of the bed defines the pitch face of the blade, while the lower surface of the top defines the suction surface. The centre line and the approximate shape of the blade are marked out on the moulding site. Wooden shuttering is erected to form a box into which the sand–cement mixture is rammed, together with iron reinforcing bars. The pitch face is then formed by a process known as 'strickling', using a board with a long arm at one end and a roller at the other. The roller runs on an inclined rail, set so that the surface of the blade form can be generated.

Next, the blade pattern is constructed and the top of the mould made, and left to harden. Each of the blade forms is made in turn. All the mould surfaces are cleaned, then the top parts of the mould, substantially reinforced, are placed in position and secured by T-bolts. *Plate 8* shows a mould in course of manufacture. The total weight of a mould for such a large propeller is about 300 tonnes. For some hours before pouring, hot air is blown into the mould to remove all moisture from the mould surfaces; in the case of the propeller illustrated, this period was 48 hours.

The nickel–iron–aluminium bronze is melted in large electric induction furnaces; the temperature is brought to 1150°C in the ladles and 85 tonnes of molten metal is poured into the mould through three openings, known in the foundry trade as 'sprues'. The pouring time is of the order of thirteen minutes; after the main bulk has been poured the casting is 'fed', eight tonnes of metal being added during the eight hours after casting.

A few days after pouring, the mould is dismantled and the casting lifted out and trimmed. It is then machined in the bore and on the contours of blades. The propeller is ground and polished to a smooth finish. The blade edges are machined to the designed shape and, finally, the balance of the propeller is checked and necessary final adjustments made.

Such casting operations as this make one appreciate the motto on the coat of arms of the University of Birmingham Metallurgical Society: 'The hand that wields the ladle rules the world.'

DIECASTING

Anyone who learns about foundry casting may remark that, though the pattern is used repeatedly, it is a pity that the sand mould has to be made over and over again. This handicap is overcome by diecasting, whereby permanent metal dies are used for making large quantities of castings in non-ferrous alloys. In one process the two halves of the die are made of steel or cast iron; the molten metal is poured into the die cavity, either manually or from an automatically operating ladle. The process is called gravity diecasting,* because the metal enters the die under its own

---

* In the U.S.A., gravity diecasting is known as permanent mold casting; pressure diecasting is known as die casting.

weight. In line with other modern developments, gravity diecasting has been mechanized, so that the opening and closing of the die, the operation of cores and sometimes the preparation of the die for the next cast are done automatically, especially for large castings required in considerable quantity.

At the beginning of the twentieth century, pressure diecasting was developed and today it is one of the most versatile ways of mass producing castings of great accuracy and good surface appearance. One die half is fixed to the body of the machine and the other half can be moved backwards and forwards on tie bars so that it is closed for the casting operation and then immediately opened. Molten alloy is injected into the die cavity under high pressure, causing a precise reproduction of the form of the component, which has been machined into it. Also, and very importantly, a sufficient hydraulic pressure is exerted on the closed die, so that it will not open during the fraction of a second that the molten alloy is being injected at the speed of an express train. The die is provided with a number of cooling channels, through which water or oil circulates; each portion of the die is cooled at an appropriate rate so that the casting cools rapidly and uniformly. After the metal has been injected the die is opened, the design of the component being arranged so that it adheres to the moving half of the die, furthest from where the metal entered. The casting is pushed away from the face of the moving die half by ejectors, which are steel rods 3–10 mm in diameter, disposed in suitable position in the die block. Then the diecasting can be removed either by a robot, or by arranging that it falls into a bath of cooling water, from which a conveyor takes the diecasting to the next operation.

For alloys of low melting point, especially those of zinc, the molten metal is held in a container which is embodied in the machine. A plunger, permanently immersed in the molten metal, is forced downwards and this causes a 'shot' of alloy to be injected into the die cavity. Such a machine is called a hot chamber diecasting machine.

Alloys of aluminium and copper, which have higher melting points than that of zinc, are diecast in cold chamber machines. The molten alloy is held in a crucible or other melting unit adjacent to the machine. An amount sufficient for one shot is ladled, generally automatically, into the plunger cylinder of the diecasting machine. A hydraulic ram then forces the metal into the die cavity. Magnesium alloys can be diecast in either hot chamber or cold chamber machines.

Zinc alloy diecastings are made automatically, on hot chamber machines, at speeds depending on the weight and complexity of the component. A casting weighing several kilograms may be produced at the rate of two or three per minute and a motor-car door handle at about five per minute, though this output is often increased by making a number of parts in the same die. Very small parts such as umbrella ferrules and zip-fastener slides are produced at several hundred per minute; one process engineer supervises a group of machines.

The development of automatic cold chamber machines for aluminium alloys is now well established, thanks to the use of robot mechanisms and improved methods of lubricating the die. The molten aluminium alloy is transferred automatically from the crucible to the injection mechanism of the machine; this is done by devices ranging from mechanized ladles to electromagnetic pumps.

For forty years or more gravity diecasting has been used, not only for medium-sized castings, but for large aluminium alloy components weighing up to 300 kg. So far, pressure diecastings have not reached this magnitude but the diecast automobile cylinder block, discussed on page 166, indicates the shape of things to come. A cylinder block die weighs as much as 30 tonnes, the machine is over 15 m in length and of a size which dwarfs the operator. At the moment of injection of the molten metal the die halves must be held firmly together; immense locking forces are required for large components such as cylinder blocks. Some American, European, and Japanese machines have a locking power of over 3500 tonnes, capable of mass producing complex aluminium alloy components weighing over 20 kg.

During the past five years the automation of diecasting has been assisted by the introduction of new technology, including the use of computers to obtain ideal conditions so that the molten metal will be injected into the die at the most efficient position and at optimum speed and pressure. When the size and weight of the component have been measured, the computer will 'advise' whether a particular machine will be suitable to produce it and will then give an estimate of the cost and of the parameters which will lead to the correct design of the die. In the not too distant future such computers will be able to produce the die design without reference to the drawing board.

A process known as low pressure diecasting has been developed and is proving successful and economic for producing fairly large components, such as automobile wheels in special aluminium alloys. The die is held above the molten aluminium alloy and is connected to the crucible by a ceramic tube. Air pressure applied to the surface of the metal causes a metered amount to rise up the tube and fill the die cavity.

CASTING RODIN'S 'THINKER'

Auguste Rodin's sculptures were first shaped in clay and then cast in bronze as hollow shells. The clay model would be divided into a front and a back half by inserting a line of 'shims' into the surface of the clay. One half of the model was then coated with liquid plaster about 30 mm thick; when it had hardened the shims were removed and the leading edge of the plaster treated with varnish so that it would not stick to the next application of liquid plaster, which was then put on to the second half of the clay model. When the plaster had hardened off, the two plaster shells were separated and pulled off the clay along the line formerly made by the shims. *Fig. 23a* indicates the shim line for one of the hands of The Thinker.

All of the clay was then removed from the interior surfaces of the two plaster shells. These two halves were then firmly fixed together in their original position, making a cavity which conformed in every feature to the original model. The surface was then given a dressing of liquid soap to make sure that plaster, which was now poured into the cavity and allowed to harden, would easily be separated from the plaster shells which were chipped away. The stage had now been reached where there was a plaster replica of the model (*Fig. 23b*).

Next a series of 'piece moulds' were made around the plaster replica. Rodin used plaster for these but modern foundries use a rubbery composition. In areas of 'undercutting', for example under The Thinker's arm, multiple small pieces had to be made. Each piece was detached from the plaster model and all the pieces fitted together like a three-dimensional jigsaw puzzle, so once again a cavity had been created. Molten wax was poured or brushed into the cavity so that the interior was covered with a layer of wax about 10 mm thick. Before this was hardening, metal pins were pushed through the surface of the wax and joined into the material as it hardened. The hollow interior was filled with a refractory mixture of silica sand and plaster.

We now have an interior core of refractory covered by a skin of wax; this was next covered in a block of refractory mould material as shown in *Fig. 23c*, positioned so that the wax pieces pointed downwards, and was then heated. The wax melts away leaving a cavity mould in refractory material, with a refractory core, ready to receive the molten bronze. As will be seen from *Fig. 23c*, channels were made in the mould so that the molten bronze would run correctly and fill the cavity. The system is arranged so that as the metal rises up the mould cavity it drives

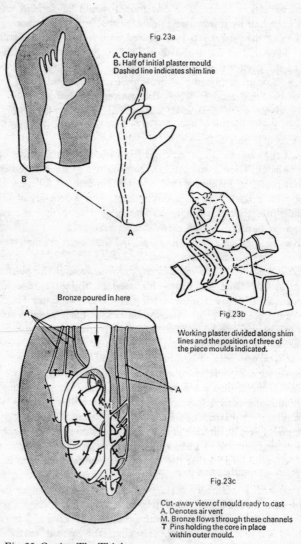

Fig.23a

A. Clay hand
B. Half of initial plaster mould
Dashed line indicates shim line

Fig.23b

Working plaster divided along shim
lines and the position of three of
the piece moulds indicated.

Bronze poured in here

A

A

Fig.23c

Cut-away view of mould ready to cast
A. Denotes air vent
M. Bronze flows through these channels
T Pins holding the core in place
 within outer mould.

Fig. 23. Casting The Thinker

entrapped air out, to prevent faults due to air-locks in the casting. The bronze is allowed to cool, the mould is broken and the casting removed. It is then trimmed and treated on the surface to provide the texture and colour required by the sculptor. The equally famous statue of Eros, cast in aluminium and illustrated in *Plate 14*, was produced in much the same way as The Thinker.

CENTRIFUGAL CASTING

For casting pipes and similar shapes, a permanent metallic cylindrical mould, without any cores, is spun at high speed and liquid metal poured into it, so that centrifugal force flings the metal to the face of the mould, thus producing a cast hollow cylinder of uniform wall thickness. This process yields a product having a dense, uniform outer surface; consequently a drain pipe or cylinder liner cast by this method is considered superior to similar ones cast in sand moulds. Cast-iron piston rings are cut from such cylindrical shapes made by centrifugal castings, and the process is now being used to make complicated components.

INVESTMENT, OR LOST WAX, CASTING

In ancient Egypt, before the pyramids were built, craftsmen produced 'lost wax' castings of astonishing beauty and detail in gold, silver and bronze. In modern times the process has been adapted and mechanized, to play a very significant role in aerospace, defence equipment and precision engineering industries. The essential feature is the use of an expendable pattern, generally of wax. A die, made of aluminium alloy, is filled with wax injected under pressure, to give a replica of the finished casting. Moving parts are incorporated in the die, to allow removal of the wax pattern in one piece. Depending on the size of the finished product, a number of the patterns are assembled; for small parts there may be several hundred patterns per assembly. For example in jewellery manufacture high quality rings are cast with about twelve impressions in each mould, while for cheaper rings over a hundred at a time are cast. The buckle of a nurse's uniform will probably be an investment casting made from a mould containing ten impressions. The largest castings, up to about 600 mm, will be made with only one pattern per mould.

The material for the runner system is added, the complete assembly is dipped or 'invested' in a high grade ceramic slurry and the wet surface is coated with fine refractory particles. This important operation forms the mould coating, which will later come into contact with the cast molten metal. After hardening the coating, the operations are repeated several times on mechanized equipment, using coarser grades of refractory, until a mould is built up of sufficient thickness to withstand the force of the molten metal during the pouring operation.

After drying, the mould is placed in a steam autoclave in which the wax is melted out – hence the term 'lost wax'. The mould is fired at about 1000°C for several hours to strengthen it and to burn off any remaining wax. After solidification of the castings the mould material is broken away from the cast metal and the runner systems are cut off.

In some ways investment casting has similarities to diecasting because large quantities are specially economic and each product has to be designed carefully to get the best out of the process. Investment castings do not have the disadvantage of joint lines which are inevitable in castings made from permanent metal moulds. The process is widely used for precision castings in stainless and other alloy steels, brass, bronze, aluminium, nickel and cobalt alloys, including those used for magnets. Gold rings and a large proportion of costume and exquisite jewellery are investment cast, coupled with centrifuging during the casting process.

CONTINUOUS CASTING

Almost all metals and alloys must be cast into a suitable shape prior to working. Rectangular slabs are the first stage for flat products such as strip and sheet; square sectional long billets for rolling into rod and wire; cylindrical shapes for extruding into rod or tube. Until the mid 1930s all such shapes were made by casting into ingot moulds having cavities of the required form. About that time the need arose for producing aluminium alloys in a fine crystalline condition. This requirement led to the development of continuous casting, which has now grown into an immense industry. First the basic principles of continuous casting of non-ferrous metals will be described.

A water-cooled copper or aluminium mould about 150–250 mm deep, having the shape of the desired cross-section of billet or slab shape, is sealed at the bottom by a retractable base (*Fig. 24*). Molten metal is poured into the cavity continuously while the base plate is slowly lowered; the metal solidifies first as a shell adjacent to the mould and solidification progresses into the centre of the cavity. The process is hastened by additional cooling, by water jets or sprays and by withdrawing the billet into a tank of water below the mould. Almost all the aluminium alloys are now semi-continuously cast; a length about 3–6

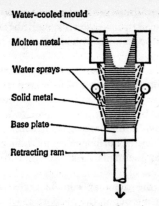

Water-cooled mould
Molten metal
Water sprays
Solid metal
Base plate
Retracting ram

Fig. 24. The principle of continuous casting aluminium

metres is poured, casting is stopped, and the solid metal withdrawn. Casting is recommenced by returning the base to its original starting position.

This development was relatively easy for metals melting at up to 700°C, such as aluminium, but it is only in the last twenty years that the metallurgical and economic advantages of continuous casting of iron and copper base alloys have been realized and today the continuous casting of steel is a large and flourishing industry, with a present output of over 400 million tonnes per annum. It is expected that at the end of the century, about 90 per cent of all the steel that is made will be continuous cast. The methods used for steel are discussed on page 125 and illustrated in *Figs. 41* and *42*, with a picture of a modern plant shown in *Plate 12*.

THE SHAPING OF SOLID METAL

If a bar of cold steel is hammered, a great amount of energy is needed to change its shape permanently. The same metal, when heated to bright redness, is softer and more pliable. The village blacksmith makes use of this and, with his hammer and forge, shapes hot steel into horseshoes or repairs tractors.

With the coming of the machine age, mechanical hammers were devised and made which would forge larger shapes than even the mighty smith was capable of tackling. One old machine is the 'tilt hammer', which can be seen at the Abbeydale industrial hamlet in Sheffield. An iron hammer head, weighing about 50 kg, is fixed at the end of a long wooden arm, pivoted at the centre. A cam rotates beneath the arm, so that the head rises and then drops by its own weight on to the piece of hot metal held beneath it.

A development of the same principle is the drop hammer, where a heavy steel die block, working between two vertical guides, is mechanically lifted about $1\frac{1}{2}$ metres above the anvil and allowed to fall under its own weight on to the metal to be forged. Two halves of a die are made; one block is bolted to the anvil at the base of the machine and the other is fixed on to the drop hammer. A hot bar of metal is firmly held on the anvil by tongs and the hammer falls so as to forge the metal between the two halves of the die.

For a fairly complicated part, the die may contain three pairs of impressions of progressively increasing detail, the purpose of which is to change the shape of the metal in gradual stages. Thus three forging operations are performed with the same die and for each of these the workman holds the piece of metal successively in the three positions during three successive blows of the hammer. The first impression causes the metal to assume a form roughly approaching that of the finished article, the next brings the metal practically to the desired shape, and the final die impression makes the forging accurate in dimensions.

A modification of drop-forging employs mechanical or hydraulic force to push the hammer downwards, thus increasing the power of the blow and the speed of the operation. Automation features in the mass-production of forgings; in modern plants several hundred can be made each hour. At the forging head, water-cooled punches and dies perform up to four operations simultaneously, producing flash-free components, often requiring no further finish-machining.

In the methods described above the hammer blow is rapid; this would not be desirable for the shaping of a large component. If, for example, the crankshaft of a car, truck or diesel engine, or a marine propeller shaft were forged in that way, the outside of the metal would receive the blow but the effect would not be transmitted completely to the interior, which would then be coarser-grained and therefore weaker than the outside. For heavy forgings an immensely powerful hydraulic press squeezes the metal with a force sometimes exceeding 10 000 tonnes. Although the hydraulic press is a more expensive piece of equipment than the drop-forge, it gives greater strength and more uniform structure to large components and it operates with less noise and vibration. *Plate 9* shows a 6000 tonne hydraulic press, producing a diesel engine crankshaft weighing about 60 kilograms.

EXTRUSION

The process of extrusion involves a similar principle to that of squeezing toothpaste from a tube or making cake decorations by squirting icing from a bag. A prodigious pressure would be necessary to extrude most of the common metals while cold, and the plant is generally designed to extrude metals hot, for example, at a cherry-red heat (700–800°C) for copper and brass or at between 400°C and 500°C for aluminium alloys.

An extrusion press of average size will take a piece of cast brass in the form of a cylinder about 120 mm in diameter and 700 mm long, weighing about 60 kg. This is heated to 750°C in a furnace and then placed in line with the container of the press. A ram pushes the metal into the press and, under a high pressure, a rod of hot metal emerges through a die at the end of the container. Thus, within the space of fifteen seconds the metal block is extruded into a rod 30 mm in diameter and 10 metres long.

The principle of the type of machine described above is illustrated in *Fig. 25a*; the process is known as 'direct' extrusion. Another method, known as the 'indirect' process, is illustrated in *Fig. 25b*. This has some technical advantages, such as less power required and greater uniformity of structure of the metal, but it involves a more intricate design of machine. The main difference between the two methods is that in the direct process the metal is forced through the die, while in the indirect process the die is forced through the metal.

Die orifices are made in a variety of different shapes to produce different sections such as curtain rails, windscreen and window sections for automobiles, and even gear wheels for small machinery, in which case the continuous gear-shaped length is first extruded and is later sectioned into gears. The process is so economical that for copper, aluminium, magnesium and lead alloys extrusion is a normal production procedure for such products as rods, bars, tubes, and strips. The extruded rod or section is sometimes finally drawn through another die in the cold state, in much the same way as for wire drawing (*Fig. 29*). This is in order to improve the dimensional accuracy and in some cases to increase the strength of the alloy.

Extrusion forms a useful starting-point for the production of tubes and pipes, since by a modification of the press it is possible to arrange for the metal to be squeezed between a solid steel mandrel and a die as shown in *Fig. 25c*. As an example of the adaptability of the process lead tubing is extruded and shrunk directly on to finished cotton-covered insulated wire, thus forming lead-covered cable for underground telephone communication.

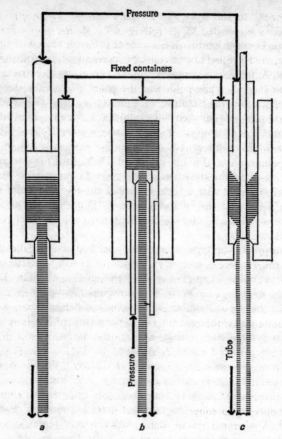

Fig. 25. Three types of extrusion

THE ROLLING OF METALS

As early as the sixteenth century soft metals such as lead, gold and silver had been rolled into sheets, but it was not until the eighteenth century that the process was employed on any large scale and for the rolling of harder metals. There are records of a rolling mill in Birmingham in 1755, capable of hot-rolling bars of iron 7·5 cm wide and reducing them to a quarter of their former thickness, with corresponding extension of length. There is little doubt that the rolls themselves and the housings of the mills were made of that useful metal, cast iron, which, thanks to the work of Abraham Darby, was being produced cheaply on a big scale. Many rolling mills of that time obtained the required power from water-wheels.

Metals may be rolled hot or cold. The advantage of hot rolling is that the metal can be reduced in thickness much more easily than when cold rolled. On the other hand the surface, finish and accuracy are not so good as those obtainable by cold rolling. Hot rolling, therefore, is for 'breaking down' large ingots; cold rolling is used to make smooth and accurate thin sheets of metal. There have been spectacular developments in the techniques and equipment for rolling metals. Although, from a tonnage point of view, the rolling of steel is by far the biggest part of the industry, the cold rolling of aluminium shows a high degree of automation, coupled with great accuracy of the product. *Plate 10* shows a modern, fast-operating aluminium strip mill combining computer-controlled automatic gauge setting and strip shape, also computer-controlled.

The end products produced by rolling are so diverse that it is difficult to give a general description, but a large works might produce steel rods in the following way. Ingots of steel weighing about 8 tonnes come from an adjacent steel works; they are put into a 'soaking pit' furnace to bring them to an intense white heat and to ensure that the temperature of the metal is uniform. At the next stage a pair of heavy rolls, driven by a 6000 h.p. motor in a 'cogging mill', have a number of grooves; as the steel is passed backwards and forwards through the rolls it travels through a different pair of grooves at each pass. A set of rolls is called a stand; often there are several stands, forming a mill train.

The rolled ingot, now in the form of a long thick bar, is reheated and further reduced in a roughing mill and then in a finishing mill. These latter two are operated by one motor, of about 4500 h.p., and are end to end, so that the one power unit can run them both; it will be noticed, however, that the power required is enormous. The finished products are long lengths of rod, the steel having been sheared during several stages of rolling.

A great amount of steel is rolled into beams for industrial buildings and bridges. The well-known I-section is first rolled to a 'dog bone' shape; a subsequent set of four rolls forms the shape of the web and flattens the flanges so that the required section is formed.

Universal beam mills have been developed to make column sections and beams accurately and competitively. Their operation is based on the principle of four-roll contact – two vertical and two horizontal. The beam is passed to and fro between these rolls which shape the web and flange. A separate stand with horizontal rolls works the edges of the flanges. The two work in tandem, doing both the roughing and finishing rolling on an ingot of steel which has previously been shaped on a cogging mill.

Other high-speed mills roll steel into rods for nails, screws and wire for fencing and suspension bridges. Enormous lengths are produced from many stands of rolls, sometimes up to twenty in number, with their rolling speeds so matched that the rod zips through the final stages with the speed of an express train.

Because of the very large amounts of steel sheets required for motor-car bodies, metal cans, galvanized sheet and many other requirements, rolling mills to make sheet in bulk have been developed to a high degree of mechanization. White-hot ingots of steel weighing 25 to 50 tonnes are first flattened in large reversing mills, to make slabs from 125 to 175 cm wide. Then the slabs, while still hot and 10 cm thick, are rolled in a series of four-high mills. These successive stands of rolls are usually arranged in twos or threes; the slab is passed through, continuously reducing the thickness. The enormous length of strip, now as little as 5 mm thick, is immediately coiled hot, at speeds of 300 to 700 metres per minute.

The hot-rolled coils are then reduced further by cold rolling. Most cold-rolling mills producing steel strip and sheet are known as tandem mills, which can be 2, 3, 4 or 5 stands, at about 4 metres apart. In order to provide the required accuracy in the finished product the working rolls are backed up top and bottom by other rolls at least three times the diameter of the work rolls; their function is to support the work rolls, preventing them from distorting. As the metal is progressively reduced in thickness the speed of the rolls must be accurately controlled to avoid buckling or breaking of the steel. This is done by small 'tension rolls' over which the strip passes between stands. If there is any tendency to buckling, the tension roll records the drop in pressure and the setting of the work rolls is rapidly corrected. The loading or squeezing power of the rolls, the adjustments for controlling the thickness of the strip, the lubrication, the mechanism which directs the metal from one stand or mill train to the next, are controlled automatically. The rolls, their housings and bearings, are designed so that in case of breakdown rapid replacement is possible. The work rolls, weighing a few tonnes, are changed regularly, requiring about 15 minutes. Occasionally, when the backing rolls and housings need changing, this involves moving over 50 tonnes.

The power required for cold rolling is greater than that for hot and the requirements of accuracy and surface finish are more precise. One cold-rolling mill in the U.S.A. accepts strip up to 5 mm thick and from 100 to 180 mm wide, weighing up to 40 tonnes, and cold-rolls it down to 0·15 mm thick. The mill has five sets of rolls in tandem and as the strip passes from one to the other it gathers speed till the last set of rolls is delivering sheet at 1500 metres per minute. All the sequences of operations, the tracking of the coil of metal, and the regulation of the rolls are computer-controlled, while closed-circuit T V is used to assist supervision.

As cold steel is harder than hot steel, cold steel mills need harder rolls and, as mentioned above, the work rolls have to be supported to prevent distortion. In the Sendzimir cold mill, illustrated in *Fig. 26*, the tungsten carbide work rolls are backed up by a cluster of heavier rolls. This type of mill is in wide use, especially for rolling stainless steel. One advantage is that roll changes take only a few seconds since the small, hard work rolls can be withdrawn quickly from this position and replaced by newly reground rolls. As in all cold-rolling mills the Sendzimir rolls are flooded with a coolant while in operation, to prevent heat from the tremendous surface pressure leading to damage of the rolls and the strip.

TUBE-MAKING

Just over a hundred years ago, steel tubes began to be used in making bicycles. The Singer *Xtraordinary* and the *Kangaroo*, both penny-farthings, built in 1878, contained steel tubes to join the handlebars to the small rear wheel. The great modern tube-making industry derived from these beginnings and, though the products of the tube-makers' skills enter into a thousand industries, a close connection has been retained with cycle-making. Nowadays the welded steel tubes of bicycles are shaped and manipulated, to be thick where the stresses are greatest and thin where the stresses are least, so that they possess the correct combination of strength, flexibility and lightness. After cycle manufacturing demonstrated the value of tube construction, their use spread to motor cycles; now tubes feature in products ranging from hypodermic needles (famous as the smallest tubes ever made) to land and marine boilers, hydraulic piping, axles of cars, trucks and railway wagons, and to many of the things we use at home or in sport – perambulators, furniture, javelins and millions of golf-club shafts.

One of the world's largest manufacturers has modestly described the two basic ways of making tubes as either wrapping an accurate hole in steel or pushing a very strong hole through a steel bar. The first method was used for making lead pipes two thousand years ago and today a

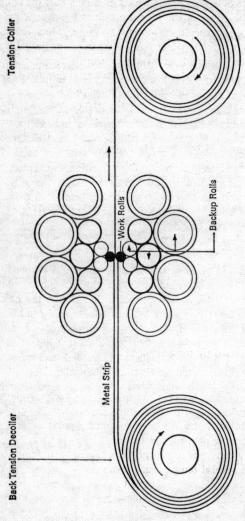

Fig. 26. Rolling of strip in a Sendzimir mill

modern derivation of that process makes untold metres of electric resistance-welded tubes.

The process begins with coils of steel strip which are fed automatically through a series of forming rolls which gradually curl the flat strip into a tube shape. The edges of the tube are heated by electric induction and pressed together so that they join by welding. Then the tube travels through a series of rolls which size and straighten it. Although many such tubes are supplied in straight lengths, a great number are manipulated – bent, tapered, flanged, swaged, screwed, tapped and slotted – into the shapes required by manufacturers.

The second type of process, where the 'strong hole is pushed through a steel bar', has many variations but the products are generally described as seamless tubes. As long ago as 1890 the Ehrhardt process was patented in Britain and proved ideal for producing tubular items which required a good solid strong end, such as high-pressure gas cylinders. This process was developed and automated until today millions of gas cylinders have been made under quality control conditions of the utmost severity, aided by ultrasonic testing and continuous appraisal throughout the process, starting with the chromium–molybdenum alloy steel bar and finishing with the gas cylinder. A permanent record is kept of every cylinder made in the last seventy years. *Fig. 27* illustrates the stages in the manufacture of a typical high-pressure gas cylinder.

The Mannesmann process for making seamless steel tubes is illustrated in *Fig. 28*. A solid rod of hot steel is spun between two mutually inclined rolls which rotate in the same direction, so that the rod is pulled forward between them, and passes over the mandrel which is shown in the drawing. In this way a thick-walled tube is produced, the dimensions of which can be varied by the setting of the rolls and the size of the nose-piece of the mandrel. After that the tube-shape can be fabricated into a thin-walled tube by further processes.

A derivation of the Mannesmann process has been developed, using a piercer with three rolls, which provide greater accuracy and concentricity than was possible with the Mannesmann mill. Round steel bars are cut to length by automatic oxypropane burners. Each piece is check-weighed and passed by a conveyor to a large furnace with a rotating hearth which passes through several temperature zones in the furnace so that finally the steel has been heated to 1250°C. The hot billet is passed to a hydraulic press which makes a cone-shaped indentation in the centre of the billet's end, making it ready for the piercing operation which will follow.

Next, each billet passes through a three-roll piercer where three

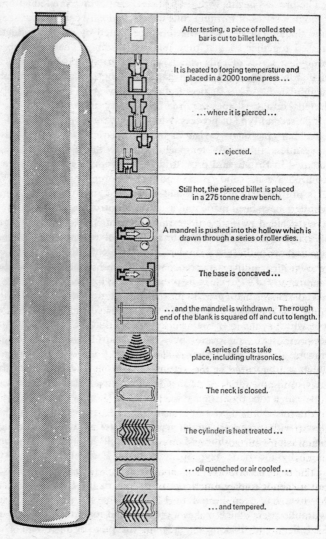

After testing, a piece of rolled steel bar is cut to billet length.

It is heated to forging temperature and placed in a 2000 tonne press ...

... where it is pierced ...

... ejected.

Still hot, the pierced billet is placed in a 275 tonne draw bench.

A mandrel is pushed into the hollow which is drawn through a series of roller dies.

The base is concaved ...

... and the mandrel is withdrawn. The rough end of the blank is squared off and cut to length.

A series of tests take place, including ultrasonics.

The neck is closed.

The cylinder is heat treated ...

... oil quenched or air cooled ...

... and tempered.

Fig. 27. The making of a high-pressure gas cylinder
(*Courtesy T I Chesterfield Ltd*)

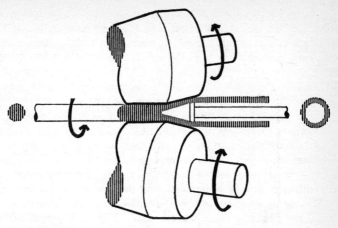

Fig. 28. Mannesmann tube-making process

barrel-shaped rolls grip the billet and force it over a conical plug in a similar, but more controlled, way than that described for the Mannesmann two-roll process. This produces a thick-walled short tube, known in the tube-making industry as a 'bloom', which must then be converted into a long tube of the required wall thickness and bore. A mandrel bar is fed into the bloom and both are passed through a series of eight stands, each consisting of a pair of grooved rolls. Each succeeding pair of rolls reduces the diameter and thickness of the bloom, while the bore is supported by the mandrel. A typical tube formed at this stage of the process would be 20–25 metres long. The next stage converts this into a finished tube three or four times as long on a 'stretch reducing mill', consisting of a series of 24 stands, each with a cluster of three grooved rolls which progressively reduce the diameter while increasing the length of the tubes. The speed of each roll cluster is set individually to enable the stretching action to take place between each stage. The tubes are examined for surface defects by visual and magnetic crack-detection techniques.

The basic description given above traces the stages in tube manufacture but it cannot convey much idea of the vast scale of production. This can be pictured by thinking of the large tonnage needed to satisfy all the industries that require tubes and by the speed of operation, metal tubes nearly a hundred metres long coming from the reducing mill at $4\frac{1}{2}$ metres per second and then being automatically cut to the required lengths.

Extrusion, which has been discussed on page 79, is used for making tubes in lead, copper, magnesium and aluminium, the process being based on the principle illustrated in *Fig. 25c*. A hydraulically operated extrusion press of 3000 tonnes' capacity is used to make stainless steel tubes with bores from 25 to 165 mm and outside diameters of up to 190 mm.

The processes so far described are for making tubes from hot metal. If a tube of optimum surface appearance and strength is required for, say, aircraft structures or hypodermic needles, a further process of cold drawing is applied. With this method the tube, pointed at one end, is then inserted through a die and gripped by a movable drawbench carriage which pulls the tube through the die and over a plug which has been screwed into position on a mandrel. This has the effect of reducing the diameter and the wall thickness of the tube and increasing its length. The cold-drawing process is used, not only for ordinary steels, but also, for example, high-strength tubes such as those of manganese–molybdenum alloy steel for racing cars and cycles, and in specialized equipment such as mountain rescue cradles.

THE DRAWING OF WIRES AND RODS

Wires are produced as shown in *Fig. 29* and are always made at a comparatively low working temperature. The raw material for making wire is hot-rolled rod about 6 mm diameter. After annealing, cooling and removal of scale, the steel rod is pointed at one end and then inserted through a hole in the die slightly smaller than the size of the rod. The pointed end is gripped

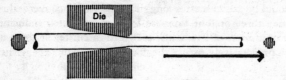

Fig. 29. Wire drawing

from the other side and the rod pulled through, thus reducing its diameter. In modern continuous methods, particularly applied to copper, the wire is drawn through a first die, turned round a roller, then passed through a second die, and the same procedure is repeated several times, using successively smaller holes and rollers of increasing peripheral speed. By the time the wire enters the last die it is much reduced in diameter and is travelling at high speed. In order to save continual rethreading of the machine, lengths of rod are welded together before drawing.

SHEET METAL WORKING

Metal parts made from strip and sheet offer advantages of less weight and lower cost when compared with castings and forgings. The mass production of sheet metal can be traced to the middle of the nineteenth century when sheets of tinplate were stamped to form food containers. The rapid growth of this important industry came when automobiles began to be mass produced and domestic equipment, from toasters to washing machines, started to be required universally.

Low carbon steel is the principal raw material. The process begins with a blanking operation in which huge coils of metal strip, often weighing thirty tonnes, are automatically unrolled and cut to the required size by shearing dies. Then the blanks are stamped in a pair of metal dies which have been shaped three-dimensionally to the required form and which are mounted on great hydraulically or mechanically operated stamping presses. The next step is flanging, in which the metal is bent to provide the areas which will be used to fasten the stamping to other components. This is often followed by a final stamping operation to sharpen the contours or corners.

The modern stamping line is highly automated, with mechanical fingers for feeding and transferring the metal, lubricant sprayers synchronized with the operation of the press, conveyor systems and an automated inspection system which functions without holding up the sequence of the manufacture. Some very large machines known as transfer presses will make complete motor-car wheels in about five sequential operations from coiled strip.

OTHER METHODS OF SHAPING METALS

A button such as is sewn on uniforms is made from metal strip in a number of stages; circular blanks are punched from a strip of annealed brass or nickel silver and each blank is placed in turn on the anvil of a stamping machine. One half of the die is mounted on a heavy steel block which is raised to a given height and dropped on the metal blank, thus stamping the pattern on the front of the button. The back of the button is stamped separately into the form of a shallow cup, and the shank which forms the eye is inserted. In shaping both the back and front a lip is made at the edges and the button is completed by forcing the lip of the back of the button underneath that of the front.

Collapsible toothpaste tubes and patent medicine containers are made by the process of 'impact extrusion'. This method, akin to the extrusion process described on page 79, depends on punching a small blank of cold metal in a die. The metal is squeezed between the die wall and the punch, producing a hollow, thin-walled container. Some aluminium teapots and hot water bottles are made from sheet metal by 'spinning'; nails are made from wire by a continuous cold-heading and pointing operation; cartridges are made by a 'deep drawing' operation.

COIN MAKING

The minting of coins is an important, though somewhat exclusive, branch of the shaping of metals. In Britain's Royal Mint at Llantrisant copper–nickel alloys are melted in electric furnaces and poured into casting machines from which they emerge as slabs about 13 mm thick and 200 mm wide. These are rolled into strips of the required thickness for the coins, sometimes with an intermediate softening process. Discs are punched out of the strip, then softened by heating on a belt passing through a furnace. After cleaning, the bright discs are 'struck' into coins by pressing between dies engraved with the required features and lettering.

The seven-sided 20p coin is made of an alloy of 84% copper and 16% nickel. The £1 coin, circulated from April 1983, contains 70 per cent copper, $5\frac{1}{2}$ per cent nickel and $24\frac{1}{2}$ per cent zinc. The compositions and shapes of the two coins were selected so that the coins might be uniquely distinguished in electronic vending-machines.

British coins are tested at what is known as the Trial of the Pyx. The word denotes a small metal receptacle in which the Host is reserved in the Tabernacle of Roman Catholic and some Anglican churches. Each year, out of every 15 pound troy of gold, a coin is put into the Pyx for testing at Goldsmiths' Hall in the City of London. The coins are tested for accuracy of weight and correct assay, in the presence of the Queen's Remembrancer and a jury of goldsmiths. Cupro-nickel coins are also tested for weight, composition and diameter and the verdict is delivered in the presence of the Chancellor of the Exchequer. The trial was initiated in the reign of Edward I and took place in the chapel of the Pyx in Westminster Abbey.

MACHINING OF METALS

All metals can be cut but there are great differences in their resistance to cutting; for example, it is much easier to cut an aluminium milk-bottle top than a thin razor-blade with a pair of scissors. It does not follow that a soft metal can be machined easily at high speed, by turning in a lathe. The soft metal being cut may build up on the cutting edge of the tool and interfere with further cutting. Faults in the manufacturing process may form inclusions which lead to rapid wear of the cutting tool tips. For example, if an aluminium alloy has been overheated when molten, aluminium oxide is formed, which causes hard spots in the solidified casting. These spots play havoc with cutting tools, especially in rapid automated machining processes. If an aluminium–silicon–copper alloy containing more than limited amounts of iron, manganese and chromium is 'stewed' in a crucible when molten, a heavy sludge is formed which sinks to the bottom of the crucible and, when the alloy is cast, causes hard spots to form which will cause trouble in machining.

To ensure high output and rapid machining many alloys are made with added elements which make fast machining possible. Small, rounded, low melting point constituents evenly distributed through the metal assist rapid machining. Well-known examples are steels containing lead, or manganese sulphide particles; copper and brasses containing lead, selenium or tellurium particles.

The shaping of metals on lathes, drills, millers and many other types of machine is an industry which ranges from the home workshop to the giant automated plants of the motor manufacturers. Massive forging presses shape and manipulate steel ingots weighing 30 tonnes or more and the forgings are then machined to make drum winders for coal mining, ships' propellers or equipment for steel-making plant. The machines on which such parts are brought to their final shape are correspondingly immense; for example, lathes 30 m long have to be accommodated. At the other end of the scale, tiny holes for wrist watches are drilled with laser beams to obtain the greatest possible accuracy. Ever since metals were first employed, man has used his ingenuity to devise new ways of shaping metals and there are still no signs that his technical resourcefulness in this respect is being exhausted.

9

TESTING METALS

Metals are tough; they are employed where high stresses and strains have to be endured. A rod of steel 30 mm in diameter – just over an inch – can

Steel rod 30mm diameter

Fig. 30. The strength of steel

support a load of over 25 tonnes without fracture. *Fig. 30* attempts to convey some idea of the load which could be borne by such a slender piece of steel.

Different kinds of stress may be experienced in service; for example, the hauling rope, made of stranded steel wire, holding the cage of a mine shaft, is subjected to pulling, or 'tensile' stress, while the vertical columns supporting a bridge suffer mainly compressive stress. Almost all moving parts of machinery undergo rapid changes or combinations of stress; for example, the connecting rod of a steam engine is subjected to alternations of tensile and compressive stresses, while the axle of a railway carriage suffers a combination of bending and twisting.

One of the tasks of the metallurgist is to specify suitable alloys which will endure the stresses encountered in service, and metals must be tested so that their mechanical properties, and especially their strength, can be assessed and compared. When a metal is selected for service, mechanical tests must be carried out as an inspection routine in order to make certain that, throughout the production of batches of that metal, the quality is maintained. The principal method is to test to destruction representative samples of the metal, thus subjecting them to much more severe conditions than they will normally endure. The designer estimates the likely stresses in the part he is designing and, with a knowledge of the mechanical properties of the metal, he is able to calculate the shape and size required, allowing a factor of safety.

THE NEWTON

Tensile strength is measured in force per unit area; engineers in Britain have usually stated the strength of metals as 'tons (force) per square inch', abbreviated in practice to 'tons per sq. in.'. In the U.S.A., tensile strengths are generally stated as 'pounds (force) per square inch'. This particular convention, although widely used, can lead to confusion of thought between the mass unit, the pound, and the pound force (which is the *force* exerted on a pound *mass* by gravity). A similar situation exists with the metric system of units widely used on the Continent. In this system the mass unit is the kilogram (kg) and the force unit the kilogram force (kgf) and here again the distinction is often blurred in practice.

With the development of metrication in Britain the opportunity was taken to adopt the 'Système International d'Unités', for which the abbreviation is 'S I' in all languages. This system is gaining acceptance and is being legally adopted by all major European countries and in many other parts of the world. In S I, the mass unit is the kilogram and the force unit has a distinctive name, the newton, commemorating the great scientist Sir Isaac Newton and his work on the force of gravity. The newton (N) is the force required to produce unit acceleration on a mass of one kilogram.

From the well-known equation: Force = Mass × Acceleration, it can be seen that the force required to produce standard acceleration (9·807 metres per second) on a kilogram mass is

$$\text{Force} = \text{Mass (1 kg)} \times \text{Acceleration (9·807)}$$

The force of 9·807 newtons is equal to the kilogram force in the metric system. A tensile strength previously expressed as 1 ton (force) per square inch becomes, in S I, 15·44 newtons per square millimetre. The derivation of this conversion factor can be seen from the following:

$$\frac{1 \text{ ton (force)}}{1 \text{ (inch)}^2} = \frac{1016 \text{ kilogram (force)}}{(25·4 \text{ mm})^2} = \frac{1016 \times 9·807}{645·16} = 15·44$$

Therefore one ton force per square inch equals 15·44 newtons per sq. mm. When kilograms per square millimetre are converted to newtons, the factor 9·807 is used.

In *Table 8* on page 108 we have shown mechanical strengths of various metals in the formerly used tons per sq. in. and newtons per sq. mm (N/mm²). A similar comparison was included in *Table 6* on page 61.

THE TENSILE TEST

The general principles of measuring the strength of metals are similar. A typical test piece would be taken from a rod of mild steel, about 130 mm long, turned on a lathe so that its diameter across the narrow parallel part of the section is 13·82 mm; its cross-sectional area is therefore exactly 150 sq. mm. The shape of the finished test piece is illustrated in *Fig. 31*. Two

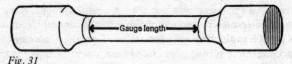

Fig. 31

marks are punched or scribed on the parallel portion, at a distance of 70 mm apart; they will be used in measuring the amount of stretch which the metal undergoes. The test piece is placed in a tensile testing machine, which is capable of applying steadily increasing and measured pulling force on the test piece and thus recording the tensile strength of the metal.

The stress is gradually applied but, though the recording dial soon shows that a 3000 kg force is being exerted, the steel appears unchanged. If, however, the distance apart of the two marks is measured with an accurate instrument, while the steel is still under tension, it is found to have increased slightly, less than a quarter of a millimetre. If now the force of 3000 kg is removed, the steel returns to its original length. In other words the metal has so far behaved elastically. The relationship between the applied stress and the amount of strain (in this case the extension of length) is known as the Modulus of Elasticity.

When the tensile force reaches about 4000 kg an important stage is reached in the process of stressing the piece of metal. This stage is known as the 'elastic limit' and indicates the maximum dead loading to which the steel can be subjected without deforming permanently. If the load were increased to over 4000 kg and again removed, the steel would not return to its original length but would remain permanently stretched. In the test we have described, the elastic limit was reached at 4000 kg; the cross-section area was 150 sq. mm; the stress therefore was about 27 kg force per sq. mm (265 N/mm²) at the elastic limit. In the elastic range, the metal, when stressed in tension, stretches only a very small amount, but after that an increase of stress causes visible extension which can be measured with a pair of dividers. At above 8000 kg the steel continues stretching and, as the load further increases, a neck gradually becomes apparent at the centre, until finally the test piece snaps at the necked portion, the broken pieces being shown in *Fig. 32.*

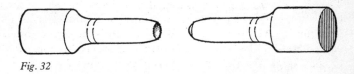

Fig. 32

The original cross-section area of the test piece was 150 sq. mm, and this area is used in the final calculation, which shows that the steel broke at about 55 kg force per sq. mm. Converting now to newtons as shown on page 94 the tensile strength was 55 × 9·807, or about 540N/mm². Had the same test been done in the 'pre-S I-unit' era, its tensile strength would have been stated as about 35 tons per sq. in.

The two broken pieces of the test bar are then fitted together and the distance measured between the two marks which were originally 70 mm apart. Owing to the stretching of the metal before the break, these marks are now found to be 87 mm apart, representing an elongation of 25 per cent. This figure is recorded as a useful guide to the ductility of the steel.

In the test which has just been described, the stressing and eventual breakage of the steel proceeded in two stages; first there was an *elastic* range in which the metal did not distort permanently under stress, then there followed a *plastic* range at higher stresses, in which the metal underwent permanent distortion. Three properties have been determined, the elastic limit, the tensile strength, and the elongation; this information can be put to practical use by an engineer or designer, as three simple illustrations may show:

1 The Elastic Limit. In designing a structure such as a bridge, it is essential to know the elastic limit of the material, for if a girder were subjected to stress above its elastic limit it would suffer a permanent dimensional change; this might increase stresses dangerously in other girders connected to it, and lead perhaps to collapse of the whole structure.
2 The Tensile Strength. A ship's hawser might be subjected to an unusually severe stress, for example when towing a crippled ship in a gale; the tensile strength indicates the greatest stress that could be applied without the metal breaking.
3 The Elongation Figure. If a metal is intended to be shaped by a deep drawing operation, it must be ductile; in other words, its elongation figure should be high (the figure given for steel we discussed above would be counted as moderately good). A metal with only a low elongation would crack when subjected to deep drawing and would be unsuitable for making, say, the body of a fire extinguisher.

OTHER MECHANICAL TESTS

Particularly in the testing of steels it has become customary to measure one other figure during the tensile testing of the metal. The 'proof stress' is the amount of stress required to cause a permanent stretch; for example, 0·07 mm on a 70 mm gauge length would be known as the 0·1 per cent proof stress. In some metals this is equivalent to the elastic limit but is easier to measure. Since the early 1970s there has been an increasing tendency to specify a 0·2 per cent proof stress for ferrous and non-ferrous metals, and eventually the 0·2 per cent proof stress will be universally specified. However, for the moment any proof stress figures quoted in this book are 0·1 per cent.

Hardness is also usually determined during the testing of metal products. Although it is such a familiar word, the meaning of hardness is difficult to define with precision, but technically it may be taken to mean resistance to deformation, and this is the basis of the usual hardness tests. A hardened steel ball or a diamond point is pressed into the prepared surface of the metal for a given time under a given load. If the metal is soft a large indentation is made, while if hard the impression is small; the area of the indentation is determined and the hardness figure calculated on the basis of load supported per unit area of the indentation. Dr Johan August Brinell, of Sweden, devised a well-known method of hardness testing, and Brinell hardness numbers are frequently used in comparing the hardness of different metals and alloys. More recent forms of hardness testers, based on similar principles, are the Rockwell machine, which was devised in America, and the Vickers diamond-pyramid hardness-tester, designed in England; in both machines a diamond, which does not deform under heavy loads, makes the indentation. The hardness of thin sheets of metal can be measured more accurately than by Brinell. Recently, a fully automatic hardness tester, Brinscan Mark II, has become available. This instrument has an automatic microscope indentation measurer and is sufficiently robust to be incorporated in production lines, carrying out up to 500 hardness tests per hour.

The Brinell hardness of some well-known metals and alloys, together with their tensile strength, proof stress and elongation figures are given in *Table 8* on page 108. Various machines use different scales of hardness but these can be correlated by the use of conversion tables. For example, the diamond-pyramid hardnesses of copper–zinc alloys are shown on *Table 6* on page 61. These can be converted to Brinell hardness by a reduction of approximately 10 per cent.

Other methods have been devised from time to time to assess the mechanical properties of metals. Many of these are simple workshop tests; for example, a strip of metal may be tested by counting the number of times it can be bent backwards and forwards over a radius until it breaks. Another more elaborate test on sheet metal consists of forcing a hemispherical plunger into a clamped piece of sheet until it just fractures; the depth of the hollow thus formed is then measured. This is known as the Erichsen test.

In the testing of that sophisticated metal structure the Severn Suspension Bridge, a simple 'wrap test' was used to verify that the galvanized wire used in the cables was satisfactory. A specimen of the wire had to be wrapped twice around a mandrel of diameter equal to three times the diameter of the wire without fracturing and without showing any flaking or cracking in the zinc coating.

Under conditions of very sudden loading or shock, some metals behave differently from what one would expect on the basis of the tensile test alone; for example, steels of certain grades possess high tensile strength but are weak under impact. On the other hand, another type of steel with a lower tensile strength might withstand a severe impact without failure. One type of testing machine was developed by Edwin G. Izod to test the behaviour of metals under impact. A notched metal test piece is broken by a heavy swinging pendulum and the amount of energy required to break it is measured. During the past twenty years the Izod test has been superseded by the Charpy impact test, which has the advantage that the test pieces are smaller and easier to machine and that the Charpy tester is more accurate than the Izod machine.

NON-DESTRUCTIVE TESTING

In most methods of mechanical testing, part of the metal sample is fractured or damaged; one can never be fully certain that the test piece is in every way typical of the bulk of the metal to which it relates. However, many millions of tests have shown that in most cases the test piece does reflect the average properties of the batch of metal being processed and hence indicates its subsequent behaviour in service.

For high-duty uses, such as in aircraft, or for nuclear power, it is important that *all* metal going into service is given a thorough and complete inspection; non-destructive testing then becomes essential. During the past fifty years, advances in the development of reliable scientific instruments have made possible the hundred per cent inspection of metal parts without destroying or damaging them. X-ray apparatus is used for 'shadow' examination of castings and wrought metal shapes, provided they are not too dense or thick in section. Thus light alloys for aircraft components are examined for porosity, and variation in grain size.

Cracks are detected by immersion in penetrating fluid of the paraffin type. After withdrawing the component from the liquid the outside surface is dried and time allowed for any liquid which has penetrated the crack to seep out again and be revealed on the surface. Improvements in this procedure have been made by putting fluorescent chemicals in the fluid and examining the part under ultra-violet light, or covering the surface of the casting with a lime wash or chalk to reveal the seeping fluid.

Where magnetic materials, such as iron and steel, are involved, a range of non-destructive tests based on their magnetic properties is used. One involves putting the part in a strong magnetic field and dusting it with fine particles of magnetic iron oxide. The iron oxide congregates where any intensive changes in magnetic intensity occur, such as in the region of cracks and, to a lesser extent, where sub-surface porosity exists.

With non-magnetic metals, such as copper and aluminium alloys, eddy-current tests may be used to detect surface flaws. This procedure involves the generation of small localized electrical high-frequency currents just beneath the surface of the tube or bar of metal under test. Search coils of copper wire pick up the residual eddy currents and compare them with a standard uniform specimen. This technique discovers whether metal tubes are free from internal defects; it is particularly difficult for the human eye to detect such faults.

The workshop test of striking a metal object with a hammer and listening to the sound, characterized by the railway wheel-tapper, has developed into examination by ultrasonic sound waves. These are in the same frequency range as those used for submarine detection but they are generated in contact with the metal surface under examination and are then transmitted through the body of the metal, reflected on the far side, and picked up again by a search crystal close to the generating crystal. The search crystal transmits a signal to a cathode ray oscillograph which reveals 'echoes' if cracks, porosity, or other cavities are present within the body of the metal. This technique is used in the aircraft industry for the examination of metal parts before they go into service, for example spar booms in aircraft wings. Die blocks for plastic injection moulding, forging and diecasting are ultrasonically tested to determine whether the steel is free from internal flaws, thus preventing the waste of man-hours that would occur if a defect in the die block were not revealed till the last machining operation.

ACOUSTIC EMISSION

Noises inaudible to the human ear can be recorded as wave forms by suitable instruments, so that we can listen to the agony of metals as they undergo stresses and strains. Such methods can be used to detect and locate deformation in metallic and non-metallic structures long before failure due to cracking would occur. In the future acoustic emission will be developed so that, for example, the conditions in a pressure vessel could be monitored to predict the time of ultimate failure and withdraw it from service before an accident.

FATIGUE FAILURE

So far, the tests which have been described indicate how sample pieces of metal behave when they are stressed *once only*, whereas in service metals often have to undergo thousands, sometimes many millions, of reversals of stress. For example, the shaft of a gas turbine aero engine may be revolving at 14 000 to 16 000 times per minute, and a journey of five hours' duration would mean that the stresses in the metal alternate or fluctuate over four million times. The turbine blades themselves also vibrate in a complex manner; frequencies of up to a million per minute have been recorded, although the stresses involved were small.

The tensile test is not necessarily a criterion of the capacity of a metal to stand up to repetitions of alternating stresses. This was realized over a hundred years ago, when bridges of wrought iron and, later, steel were replacing stone and brickwork bridges. It became essential to learn about the capacity of metals to undergo many repetitions of stress, and in the early 1860s, at the request of the Board of Trade, Sir William Fairbairn carried out some tests in which a load was raised and lowered on to a large wrought-iron girder. It was calculated that the application of a single load of 12 tons would be required to break the girder, but Fairbairn found that if a load of a little more than 3 tons were applied 3 million times the girder would break. He concluded, however, that there existed a certain maximum load, under 3 tons, which could be applied an indefinite number of times without fracture occurring. A few years later the German engineer, August Wöhler, carried the work further, and since that time the 'fatigue' of metals has been intensively studied, particularly since the First World War. Among scientists in Britain, Dr Herbert J. Gough and his fellow-workers played a prominent part in this branch of metallurgy, which is so important in our era of high-powered engines and supersonic aircraft.

The method developed by Wöhler for testing the fatigue of metals is used extensively. A specially shaped test piece is gripped at one end, while at the other end of the specimen a ball race is fitted, from which a load is suspended. The test piece is then very rapidly rotated from the gripped end by a fast revolving electric motor so that under the action of the overhung load it is alternately stressed in tension and compression once each revolution. A mechanism for counting the number of revolutions or reversals of stress, often amounting to 50 millions or more, is attached to the machine, which is kept running until the specimen breaks, perhaps after a short time. The load and the number of stress reversals are both noted; the same procedure is then carried out with a similar test piece of the same metal, but with a smaller load. This time the number of reversals of stress necessary to cause failure is greater. From a knowledge of the load and the dimensions of the test piece the amount of applied stress is calculated; in the type of test we have been considering the stress is expressed with a plus and minus sign, since there has been a measured amount of tension and an equal amount of compression. From the results of a number of such tests on steel a graph similar to *Fig. 33* is obtained.

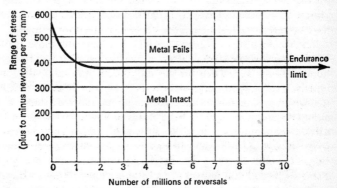

Fig. 33

It will be noticed that when the stress range is high a comparatively small number of alternations of stress are necessary to break the metal, but as the range of stress becomes lower a much greater number of reversals can be endured until, when the stress alternates from 385 newtons per sq. mm compression to the same amount of tension (the stress alternates about a mean of zero), the graph has flattened out, indicating that this range of stress could be applied an indefinite number of times.

Such a steel is said to have an 'Endurance Limit' of plus to minus 385 newtons per square millimetre ($\pm 385\text{N/mm}^2$). If a slightly higher stress of, say, $\pm 400\text{N/mm}^2$ were applied, the life of that steel specimen would be only one million reversals; parts of a motor-car may be subject to more than that number of alternations of stress within a year of normal usage. In the case of a bridge, however, heavy stresses may be applied only a few times daily, for example, every time a train passes over, although there would be many millions of smaller stresses due to wind changes; such a bridge could be designed on the basis of a few hundred thousand applications of the maximum stress.

Table 7. The endurance of an aluminium alloy under fatigue at varying ranges of stress

Stress range Newtons per sq. mm	Number of reversals endured before breakage
Plus to minus 190	1 000 000
Plus to minus 160	5 000 000
Plus to minus 150	10 000 000
Plus to minus 123	50 000 000
Plus to minus 118	100 000 000

So far, we have discussed fatigue limits of steel at normal temperatures; but at elevated temperatures, critical for each alloy and metal, there is no definite fatigue limit. Fortunately steels have a fatigue limit at room temperature, but many non-ferrous alloys do not show a definite endurance limit, but have a given life for any given stress range at ordinary temperature. Therefore, instead of the horizontal line for steel, shown in *Fig. 33*, the right-hand part of the curve would continue to slope gradually downwards. *Table 7* shows the number of reversals endured by an aluminium alloy for varying ranges of stress. It is possible to establish what is a safe load for aluminium and other non-ferrous alloys, based on a knowledge of what a part has to do in service, particularly if its anticipated length of service will elapse before the load under which it works would cause failure. Because of this, connecting rods in racing motorcycles or cars are allowed a working life of only a few hours so that the number of reversals at the high stress range will be less than the Endurance Limit. This finite life of light alloy components partly accounts for the regular overhauls that are made to aircraft after a given number of flying hours, when some parts are renewed, partly on account of wear, distortion, or corrosion, but also so that they will not be stressed a sufficient number of times to make them liable to fatigue failure.

Sometimes metal parts which are subjected to fatigue stresses in service fail prematurely under repetitions of stress lower than those established as safe by experiment. One of the likely causes of this is the influence of an abrupt change in section. One has only to recollect the tragic failure around the window frames of the cabin of the early Comet jet aircraft to appreciate the difficulty of ensuring the avoidance of design details which may contribute to low fatigue strength in the structure. In a similar manner, scratches, dents, or even inspectors' stamp marks may lead to early failure under alternating stresses. Another aspect of fatigue failure concerns the influence of corroding media such as salt water exerted at the same time as the fatigue stress; this combination of fatigue and corrosion may lower the endurance limit of metals.

CREEP

When metals such as steel are used at high temperatures under uninterrupted stresses, as, for example, furnace and steam-boiler parts, they yield very slowly, so that over a period of months or years, they stretch and may eventually fracture; this effect is generally referred to as 'creep'.

Because creep at high temperatures is most troublesome to the engineer, the mistaken notion that it is purely a high-temperature phenomenon persists. Creep can take place even at low temperatures, particularly in soft metals of relatively low melting point. Thus lead sheets for church roofing gradually thicken near the eaves.

When stress is applied to a metal under creep conditions, the following stages occur:

1 An initial instantaneous small strain, called micro-creep;
2 The primary stage of creep, during which strain or flow occurs at a decelerating rate;
3 A prolonged period, called the secondary stage of creep, during which further deformation is small and steady;
4 The creep rate accelerates, the test piece elongates rapidly and ultimately ends in fracture. This is known as the tertiary stage of creep.

The best way of determining the suitability of a metal for service at elevated temperature is by means of creep tests. A specimen similar to a tensile test piece is subjected to a constant load, surrounded by a furnace to maintain the entire test piece at the temperature. Accurate measurements are made of the increase of length, over periods of time which may be as long as one to ten years.

In many cases today metals and alloys are not required to endure prolonged exposure to high temperatures for years on end. Many aircraft engine components endure high temperatures for only 1000 to 5000 hours. This aspect has led to the development of the stress rupture test, to indicate the stress which will cause an extension at one millionth of a centimetre per centimetre of test piece per hour when exposed to a constant temperature for a period of, say, 100, 300, 600, and 1000 hours.

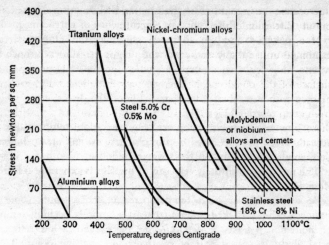

Fig. 34. Stress versus temperature for rupture in 1000 hours

By plotting these stresses a type of curve is developed as shown in *Fig. 34*. This indicates the breaking stress, for different temperatures, causing rupture after 1000 hours. The curves to the right of the diagram are in block form, to illustrate the behaviour of groups of alloys of the same family.

An aluminium alloy, for which the curve is at the left of the figure, becomes weak at quite a low temperature. The next curve represents the best obtainable from titanium alloys; it will be noticed that in the temperature range 540–650°C the titanium alloy has a creep rupture strength similar to that of steel, yet its density is only half that of steel. At higher temperatures the curves show stainless steel and some special alloys used in jet engines.

THE MECHANISM OF FAILURE

A study of the mechanical testing of metals shows that their behaviour is not always apparently consistent, and a number of questions come to mind concerning the mechanism of failure. For example, what leads to the difference in behaviour of metals when stressed (a) within the elastic limit, and (b) above it? Does the fracture of metals occur by separation along the grain boundaries or are the grains torn in half? In dealing with these questions much help has been given by studying metals under the

microscope during and after stressing and by X-ray examination of the
minute distortion of their atomic structures.

If a polished and etched metal such as aluminium or copper is
examined under the microscope while being stressed at room temperature, a widespread change appears when the elastic limit is passed. The
surface of the metal can be seen to have roughened and each grain is
marked with a number of fine parallel lines, the directions of which vary
from grain to grain (*Fig. 35b*). In some of the grains two or three series of
these lines may have developed. They are called 'slip-bands' and are
actually steps produced on the surface; they are an outward sign of the
permanent deformation of the metal grains.

The mechanism of the formation of the slip-bands can be appreciated
from *Fig. 36*. Imagine part of a grain of metal to be stressed as indicated

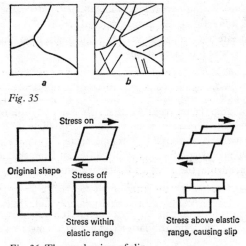

Fig. 35

Fig. 36. The mechanism of slip

by the arrows in the upper drawings. At first the grain can distort
elastically and if the stress is removed it will revert to its original form as
shown in the lower drawings on the left. But above a certain stress a part
of the grain slips like a pile of coins being pushed over slightly, and when
this occurs part of the deformation is permanent, as shown in the bottom
right-hand drawing. The planes at which the slip occurs could be
observed under the microscope in the form of slip-bands, as shown in
Fig. 35b and *Plate 24c*.

As the metal is stressed, the number and the intensity of slip-bands increase until, with sufficient distortion, the structure is so confused that the original boundaries of the grains almost lose their identity. When this state of affairs is reached the metal commences to distort in another manner, by general plastic movement.

In the early stages of deformation the slip results in the hardening of the metal. It appears probable that the slip of one block of the grain over another results in the formation of minute fragments or 'crystallites', and it is doubtful whether slip occurs again exactly in that place because these crystallites have locally strengthened and hardened the metal. The arrangement of the atoms in a grain of metal is such that certain planes exist along which slip can take place more easily than in other directions.

The research that has been done on dislocations, previously discussed on pages 62–3, has thrown a great deal of light on the reasons why metals are so strong, and how and why they fail under increased stress. It is probable that the inspiration which led to metals being strengthened by minute fibres of other materials was derived from the spread of knowledge of the effect of dislocations, and the realization that such added fibres can produce increased strength.

THE BREAKAGE OF HOT METALS

It will be remembered from page 49 that when a stressed metal is heated above a certain temperature the metal re-crystallizes. When the metal is stressed while it is above this temperature slip-bands are not observed, but a type of plastic flow (possibly different from that at room temperature) occurs from the start and often results in a considerable deformation of the metal, which usually fails by cracking around the grain boundaries. The major difference between the rupture of metals when hot and when cold is that in the former state the grain boundaries are the weakest parts, while in the latter the grains themselves are the weakest.

FRACTURE MECHANICS

Metallurgists and engineers have always paid great attention to the examination of the behaviour of metals in mechanical structures and much has been learnt from the study of components which have failed in service. In one new technique a specimen is taken from a fatigue test and pre-cracked to a known amount. The test is then continued until failure occurs; in this way the effect of stresses on the propagation of a crack can be studied. Examination of the break is then carried out with an electron-scanning microscope. Such testing has become more and more necessary because factors of safety have often been reduced to lower the cost of manufacture and therefore stress conditions in a machine or structure are more severe than before.

QUALITY ASSURANCE

The need for safety and reliability in manufactured products has led to the use of inspection routines which were applied first in the aircraft and chemical industries but which are now widely accepted. Besides working to specifications, the inspection organization and production departments are put on their honour to follow the inspection procedures and to record and mark each batch of production, so that it can be identified. The ensuring of integrity in manufacture has been developed by Rolls-Royce and by many other British concerns, as well as those in the U.S.A., Japan and other countries. Quality control groups are formed among management and workers to assure the reliability of all products going through the plant. Statistical techniques, instrumentation and sets of gauges are used; spot checks are taken from current production. The remarkable freedom from mechanical failure in aircraft provides an example of what quality assurance can achieve. For military equipment the U.K., the U.S.A., Canada and Australia use identical quality assurance procedures. Often staff members, faced with the new requirements of quality assurance, grumble about 'more red tape' but usually they agree that objectives have been achieved in a cost-effective manner, and that the status of the company has been enhanced.

CHOOSING THE RIGHT METAL FOR THE RIGHT JOB

Table 8 shows some of the mechanical properties of well-known metals and alloys. The strength has been expressed in tons per square inch and in S I units. In selecting a suitable alloy for a given purpose, designers and metallurgists should work in close co-operation, bearing in mind that a wide choice of metals and alloys of varying strengths is available. Lead has a tensile strength of only 15 newtons per sq. mm while heat-treated alloy steels are nearly a hundred times stronger.

The strength of a part increases in proportion to its effective section, but this way of increasing strength is not always desirable, as it increases the dead weight. Even in a bridge the greater part of the strength is used in supporting its own weight; increase of section augments the load which the bridge has to support. The correct procedure in designing any

Table 8. Some mechanical properties of well-known metals and alloys (These figures are approximate)

Metal or Alloy	Condition	As used for	Brinell Hardness Number	Tensile strength		0.1 per cent proof stress		Elongation per cent on 56 mm
				tons per sq. inch	newtons per sq. mm	tons per sq. inch	newtons per sq. mm	
Aluminium	Wrought and annealed	Frying pans	27	6	92	2	31	18
Aluminium alloyed with 7% magnesium	Wrought and annealed	Tubes and sheet for aircraft	80	20	309	8	124	17
Magnesium alloy, with 8% aluminium	Cast and heat-treated	Aircraft landing wheels	60	17	264	5	77	10
Copper	Wrought and annealed	Tubes	50	14	216	4	62	55
Copper	Cold-drawn	Copper wire	110	28	434	26	403	4
70/30 brass	Deep-drawn	Cartridges	160	35	540	30	463	10
70/30 copper-nickel alloy	Drawn into tube	Condenser tubes	170	38	587	30	463	8
Mild steel	Hot-rolled	Ships' plates	130	30	463	15	232	25
Alloy steel with 3·7% nickel	Forged, quenched and tempered at 400°C	Camshafts	400	86	1340	77	1188	14
0·8% chromium 0·2% carbon	Forged, quenched and tempered at 600°C	Gears	300	65	1004	57	880	22
Cast iron	Cast	Lathe beds	200	14*	216	—	—	—

* It would seem that when lawyers speak of 'a cast-iron case', they are displaying little appreciation of the strengths of metals.

component is to arrange that the metal shall be thick at just those places where strength is required; this is why girders used in building construction are flanged, for more metal is disposed at the top and bottom, where the stresses are greatest.

A metal which is ideal for enduring one kind of stress may be unsuitable for another; cast iron, for example, is strong in compression, but weak in tension. Thus it is essential to determine the type and size of the stresses to be met and then select the material most able to meet that stress or combination of stresses. A railway coupling needs to be made in a strong but ductile metal (in other words, one with a high elongation figure), so that any sudden overload can be absorbed without fracture occurring; a brittle metal would not be suitable, however strong, since the smallest stretching due to severe overload, such as might sometimes arise, would cause immediate breakage.

Not only must the metal be fit to withstand the stresses applied in service, but it must be capable of being shaped into its finished form without difficulty. So the task of the manufacturer must be envisaged, including casting, forging, pressing, machining, and heat treatment. Other special considerations of duty may have to be considered. The metal may have to work in corrosive conditions, or may require some specific magnetic or electrical properties.

The competition between high tensile steel, titanium and aluminium in racing bicycles illustrates how the function and manufacture of each component must be considered before a correct choice is made. A racing cycle designed for an endurance course which includes fast sections and mountain passes, such as the Tour de France, requires human energy to propel it along roads and up many thousands of metres; thus weight reduction is a race-winning factor. Titanium alloys, though costly, are very strong and only half the weight of steel. The Spanish rider Louis Ocana used a titanium-framed machine on some of the stages when he won the Tour de France in 1974. Unfortunately titanium tubes bend more under stress than high tensile steel; consequently to get the same rigidity the diameter and thickness of the tube have to be greater in titanium than in steel. This has two disadvantages: the weight is increased and thus some of the benefit of the change is lost, and the bigger tube causes discomfort for the well-developed thighs of the racing cyclist. To overcome this, the Irish champion Sean Kelly's titanium alloy machine, weighing only 9 kg, has oval sections instead of round tubing.

There have been difficulties in welding the highly stressed joints in titanium tubes, and a strong competitor for the manufacture of the frame is a steel containing about 0·25 per cent carbon, 1·5 per cent manganese and 0·25 per cent molybdenum. Along with this high tensile steel, titanium alloys have been used for such parts as pedal spindles, nuts and bolts. This weight-reducing combination of a carefully designed alloy steel frame and titanium bits and pieces, was an important factor in Joop Zoetmelk's victory on a Raleigh in the 1980 Tour de France.

In the meantime, French and American manufacturers have been developing the use of aluminium alloy tubes which can be joined by adhesives instead of welding or brazing. The authors of this book have passed the age when they rode racing cycles but if pressed to give their opinion on the merits of the currently competing materials, they would probably select high strength steel tube plus titanium alloy bits and pieces.

10

IRON AND STEEL

Inchtuthill, a Roman site near Perth in Scotland, was built in A.D. 83 as the advance headquarters of Agricola. The legion was withdrawn after only six years and was instructed to leave nothing that could help the enemy. Timber was removed, pottery smashed, and wattle burned; but one valuable load could not be destroyed and had to be buried in a pit two metres deep. It consisted of 763 840 small, 85 128 medium, 25 008 large and 1344 extra-large iron nails. The small ones were 50 mm long and the extra-large were magnificent tapered spikes 400 mm long, square in section, with solid heads.

In 1961 the late Sir Ian Richmond of Oxford discovered and unearthed the nails, a few of which, in the centre of the dump, were remarkably well preserved. Experts in early metallurgy reported that the composition of the metal varied from pure iron to high carbon steel. Their observations supported the view that early smelting processes produced a spongy mass of iron which was then repeatedly heated and hammered to consolidate it and to expel entrapped slag.

The Latin word for iron was *ferrum*; iron, steel and cast iron are classed as ferrous metals, indicating that they consist largely of iron and distinguishing them from the non-ferrous metals. They are at the root of our material civilization. Without them there might have been no great liners, no skyscrapers, railways, motor-cars, or tractors. If this book had been written with the space devoted to each metal proportionate to its tonnage, more than 270 pages would have been about iron and steel and only 15 pages devoted to all the other metals. As an illustration of the scale of manufacture of iron, ten large blast furnaces can produce in one year over fifteen million tonnes of cast iron; this is comparable to the whole world's annual output of aluminium. The Forth Road Bridge and the Sydney Harbour Bridge each contain over fifty thousand tonnes of steel. In Britain over ten million tonnes of iron and steel are used in our railway-line systems and two million tonnes in locomotives.

Steels may contain up to 1·5 per cent carbon, though the most widely used grades, in bridges, concrete reinforcing, car bodies and ships, contain only 0·1 to 0·25 per cent. Several elements besides carbon are present in steel; some, like manganese, have a beneficial effect; others, for example sulphur, may be harmful; steel makers reduce the amount of such impurities as much as is economically possible. The addition of nickel, chromium, molybdenum or tungsten produces 'alloy steels', including high-speed steels, stainless steels, and die steels. Cast iron contains between 5 and 10 per cent of other elements including carbon, silicon and manganese. It is produced by re-melting pig iron, which is the product of the blast furnace and is the cheapest of all metals. Alloy cast irons contain alloying elements such as nickel or chromium.

CAST IRON AND WROUGHT IRON

Before the Industrial Revolution steel was an expensive material, produced in only small quantities for such articles as swords and springs, while structural components were made of cast iron or wrought iron. Thus the first metal bridge in Europe, at Ironbridge over the River Severn (illustrated in *Plate 11*), embodied nearly 400 tons of cast iron. The project was first discussed in 1775 and work began in November 1777. In the spring of 1779, preparations were made for the erection of the ironwork. The bridge was opened on New Year's Day, 1781, and the total cost was about £5000.

Wrought iron, which had a high reputation for strength and reliability, was made by a hot and strenuous process known as 'puddling', carried out in small reverberatory furnaces in which pig iron was melted and refined by the addition of iron oxide and other substances to oxidize and remove carbon, silicon and sulphur. The temperature of the furnace was sufficient to melt pig iron (melting point about 1200°C) but as the impurities were removed and practically pure iron was obtained, with a melting point about 1500°C, the furnace temperature was not high enough to keep the metal molten. Therefore, the iron had to be rabbled into white-hot spongy lumps, bigger than footballs, which were dragged from the furnace and then hammered to squeeze out most of the slag that had been formed during the refining process. After that, the wrought iron could be hot-rolled or forged to make the required shape. It always contained a small amount of entrapped slag; *Fig. 12* on page 48 shows its appearance when seen under the microscope.

Isambard Kingdom Brunel made a historic achievement involving the use of wrought iron for the world's first ocean-going propeller-driven iron ship, SS *Great Britain*. The vessel was revolutionary in design and size; it was the first to have double-skin iron construction, five watertight transverse bulkheads, a balanced rudder and a six-blade 4-ton propeller. It weighed 2936 tons, nearly double the weight of any other vessel built at the time. The hull was constructed of overlapping wrought iron plates riveted to iron frames.

After many vicissitudes, change of ownership and alterations the *Great Britain* settled down to twenty years of voyages between Liverpool and Melbourne, including the transport of the first all-England cricket eleven to visit Australia. When her passenger-usefulness was ended, she was converted to a cargo vessel but was damaged in a gale around Cape Horn and finally beached in the Falkland Islands.

The story of the *Great Britain*'s triumphant return to the Bristol Great Western dry dock is not within the scope of this book but three coincidences must be mentioned. 19 July 1839 was the date of laying the first plates; *19 July 1843 was her launching date* and *19 July 1970* her final re-docking. Prince Albert had launched the *Great Britain* and Prince Philip was on board for her re-docking. Now it is possible to see the wrought iron which made the entire hull. Despite exposure to sea water for *over 130 years*, the plates are still more than half as thick as when they were laid down.

Fifty years after the *Great Britain* was built, Gustave Eiffel's famous tower was being erected for the 1889 Paris Exhibition. Steel was considered as a possibility but wrought iron was cheaper at that time, so the Eiffel Tower was constructed by Forges de Wendel, using about 7300 tons of wrought iron.

THE BESSEMER PROCESS

In August 1856, an Englishman, Henry Bessemer, announced his invention of a process which eventually reduced the price of steel to about a seventh of its former cost and, more important still, made it possible to produce steel in large quantities.

Bessemer proposed burning away the impurities by blowing air through molten pig iron, an idea that appeared fantastic and dangerous to the Victorian ironmasters. However, they invested large sums of money in his process, only to be scandalized that they could not make it work. Bessemer paid back all their money, spent several thousand pounds in discovering what had gone wrong, and proved that his own experiments had been with an iron containing only a small percentage of phosphorus; they had unfortunately tested his converter with cast irons of a high phosphorus content. He then tried to persuade his disillusioned clients to try his process again but they had been 'once bitten'. Bessemer then decided to go into production himself, built his own steel works in Sheffield and was soon making nearly a million tons of steel per annum. He continued to meet opposition and once, when he attempted to interest a railway engineer in the possibilities of steel for railway lines, received the reply, 'Mr Bessemer, do you wish to see me tried for man-slaughter?'

Bessemer converters, like huge concrete mixers, could be tilted to receive twenty-five to fifty tons of molten pig iron and were then brought upright for the 'blow' to take place (*Fig. 37*). Air was blown through a number of holes in the base of the converter and forced its way through the molten metal. The oxygen from the air blast combined with some of the iron, producing iron oxide which dissolved into the molten metal and then reacted with the silicon and manganese, which thus became oxidized and combined to form a slag; the carbon was removed as carbon monoxide, part of which burned and formed carbon dioxide. The time taken for the removal of impurities was about 15 minutes and the complete operation, from one tapping to the next, occupied 25 to 30 minutes. No external fuel was applied; the heat requirement was furnished by the oxidation of the impurities, and some of the metallic iron. The converter was tilted and its metallic contents poured into a large ladle. An alloy containing manganese was added while the metal was being poured into the ladle; the manganese combined with dissolved iron oxide, thus removing it from the steel. Other additions such as ferro-silicon and aluminium were made to assist the deoxidizing of the steel, then anthracite was added to bring it to the correct carbon content.

The first Bessemer converters were lined with silica bricks, known in the refractory trade as 'acid' to differentiate them from 'basic' refractories containing oxides of metals. The 'acid' Bessemer process, as the Victorian ironmasters discovered, could not eliminate phosphorus, which is harmful to steel, so low-phosphoric pig irons had to be used. In 1878, two Englishmen, Sidney Thomas and his cousin Percy Gilchrist, contributed an improvement whereby they lined the converter with 'basic' refractory bricks, containing magnesia or dolomite. Lime was added to the bath to combine with the phosphorus and silicon, and thus remove them from the iron in the form of slag containing calcium phosphate and calcium silicate. The basic lining of the converter provided conditions under which the reactions with the lime could take place without destroying the furnace lining. If silica brick were used, as in the acid process, the lime would attack it chemically.

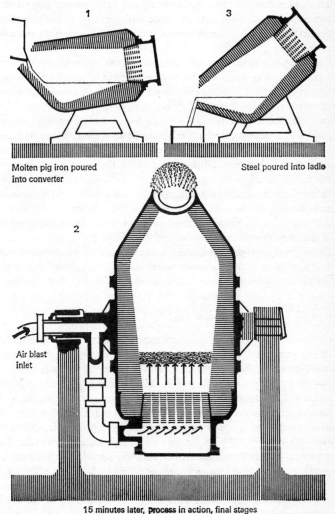

Fig. 37. The Bessemer converter

Since air, a mixture of nitrogen and oxygen, was used, the resulting steel contained nitrogen which made the metal liable to brittleness. Furthermore, the bulk of the nitrogen which was not dissolved carried away so much heat that only a pig iron of high phosphorus content could generate enough heat to maintain the required temperature of the liquid metal in the basic Bessemer process. The remedy, developed much later, was to replace the air blast by oxygen; this overcame both the difficulties at the same time, while producing steel much faster and more economically, leading to the recent revolutions in steel-making.

Twenty-three years after the introduction of the Bessemer process a tragedy occurred which caused attention to be drawn to the possibilities of construction with steel. On the night of 29 December 1879, a bridge over the River Tay collapsed while a train was crossing, and seventy-eight lives were lost. The bridge, built of wrought iron and cast iron, with 84 spans each 70 m in length, had been described as one of the engineering wonders of the world. The failure of the structure was due not so much to low strength in the wrought iron as to faulty manufacture of cast-iron columns and inadequate allowance for the stresses of wind and flood which were so great on that stormy December night. Nevertheless, the disaster stimulated engineers to reconsider the question of the materials of construction; it was realized that steel, which by that time was being produced on a considerable scale and fabricated into standard shapes, was a more suitable engineering material than either cast iron or wrought iron.

THE OPEN-HEARTH PROCESS

Dr Charles W. Siemens and his brother Frederick and later the brothers Pierre and Emil Martin of France were responsible for a method of making steel from molten pig iron and scrap in large reverberatory furnaces. They experimented with a method of 'heat regeneration', making the outgoing burnt furnace gases pre-heat the incoming gaseous fuel. By this means a sufficiently high temperature was obtained to treat large quantities of metal and to keep it molten throughout the process, enabling it to be cast into ingots when the refining was complete. The Siemens–Martin open-hearth furnace was so called because the molten metal lay in a comparatively shallow pool on the furnace hearth. *Fig. 38* shows the furnace, with the heat-regenerating chambers at a lower level.

Usually scrap steel was charged and heated previously in the furnace and molten pig iron added to it; thus the impurities in the pig iron were diluted and the refining process did not take as long as it would have

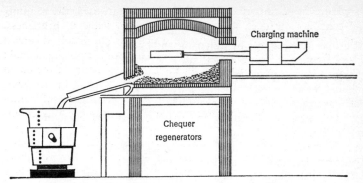

Fig. 38. Open-hearth furnace

done if the entire charge had been of pig iron. Iron oxide, in the form of iron ore or scale, was added. This, together with oxygen in the furnace gases, oxidized the impurities, the carbon being removed as carbon monoxide. The silicon and manganese were also changed into their oxides which reacted with added lime or sand to form a slag. At the end of the process, when the impurities had been brought down to the required level and the steel had been tapped, additions of ferro-manganese or ferro-silicon were made in the ladle, to bring the steel to the correct composition.

In the heyday of open-hearth production the furnaces were lined with either 'acid' or 'basic' firebricks, according to the type of pig iron employed and the grade of steel to be produced. The basic open-hearth, like the basic Bessemer process, permitted the removal of phosphorus in addition to the other impurities, whereas in the acid furnaces the refractory linings were of silica brick, making it possible to remove only carbon, silicon and manganese. As will be seen from *Table 9* on page 121 the acid open-hearths began to go out of use in the 1940s, and basic furnaces in the 1970s.

The complete cycle of operation lasted from five to fourteen hours and the open-hearth furnace of average capacity dealt with ten to twenty-five tonnes per hour. Remembering from Chapter 3 that a modern blast furnace produces 5000 tonnes of iron per day, and that some of the newest ones make twice that figure, it will be seen that one blast furnace could keep at least eight open-hearth furnaces in operation. This rather inefficient performance – compared with the enormous output of the modern blast furnace – justified the emergence of the highly productive oxygen process for steel.

THE USE OF OXYGEN IN STEEL-MAKING

The refining of steel by Bessemer and open-hearth processes removed impurities from pig iron by iron oxide and by the oxygen of the air, most of the impurities being taken into the slag. Bessemer himself had envisaged the use of oxygen but of course he could not obtain sufficient amounts, even for experimental purposes. In the 1960s steel-making took a great leap forward, thanks to the production of oxygen on such a scale that it is measured by the tonne (about 738 cubic metres at atmospheric pressure) and at a fraction of its former cost. Plants have been built adjacent to large steel-works, each capable of providing several hundred tonnes of high-purity oxygen a day.

The first developments of the oxygen process began in Austria, but soon other Continental steel-makers followed suit. They had depended to a large extent on Bessemer production of steel and it was possible for them to gain confidence in the use of oxygen by at first enriching the air blast with some oxygen, or by a mixture of steam and oxygen, or carbon dioxide and oxygen.

One of the first oxygen processes was the L–D process, an abbreviation of 'Linz Düsenverfahren', or Linz lance process. It is however often suggested that the name was derived from the initials of two separate plants in Austria, at Linz and Donawitz. The local Austrian ore was too low in phosphorus to enable the air-blown basic Bessemer method to be used, while the amount of steel scrap available was not high enough to make open-hearth steel economic. Thus a combination of circumstances in Austria provided the incentive for this major development, which radically altered the steel industry all over the world.

The L–D process consists of blowing a jet of almost pure oxygen at high pressure and travelling at supersonic speed on to the surface of molten iron, held in a converter as illustrated in *Fig. 39*. The vessel remains stationary in a vertical position throughout the blow. The speed of the reaction makes chemical analytical control essential and the installation of semi-automatic computerized 'press button' analysis machines has been an important aid.

OTHER OXYGEN PROCESSES

The development of oxygen steel-making caused a tremendous increase of productivity; in the 'pre-oxygen era' ten hours were required to make 250 tonnes of steel in an open-hearth furnace. Now it can be done in about forty minutes. During the past twenty years there have been many

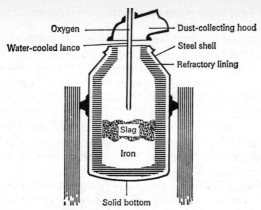

Fig. 39. L–D top-blown converter

improvements. The original L–D process was limited to the treatment of iron whose phosphorus content was under 0·3 per cent. Then French and Luxembourg steel-makers discovered that the injection of lime with the oxygen permitted the treatment of blast furnace iron of much higher phosphorus contents – up to nearly two per cent. This version of the process is known as BOP (Basic Oxygen Process).

At present the world steel-making community refines molten metal from the blast furnace, plus scrap, by blowing oxygen into the vessel and on to the bath in one of three ways, each with a number of varieties: from the top, from the bottom, or combined from both directions. The first group includes L–D and BOP, described above, the second includes the German OMB (Oxygen–Maximilianhütte–Boden) and the American Q–BOP (the Q standing for quiet, because it is less noisy and violent than the Basic Oxygen Process). There are at least fifteen combined blowing processes, in Europe, Japan, China and the U.S.S.R. One of these, known as BAP (Bath Agitated Process) is a technique developed jointly by British Steel and Royal Netherlands Steel. In normal practice interaction between the oxygen jet and the surface of the metal results in the formation of slag which is too highly oxidized. By stirring with an inert gas injected through the bottom of the furnace the composition of the steel is improved and the operation is more controllable. The oxygen is injected from the top of the converter and the stirring action is provided by argon and nitrogen blown from the bottom. The thickness of the refractory lining had to be increased from 600 to 800 millimetres, but the introduction of the Bath Agitated Process required very few other modifications of the furnaces. The whole operation is computer-controlled.

Oxygen processes present special problems of moving immense amounts of iron, scrap and liquid steel to and from the relatively small space around the furnace. Compared with its large output there is not much room for men! A single oxygen furnace can produce over three million tonnes of steel per year, which means moving six million tonnes of metal in and out.

Table 9 shows the rise and fall of the Bessemer and open-hearth processes during the years from 1870, and the emergence of oxygen steel-making since 1960 in Britain. The world recession and local problems in the British steel industry caused the dramatic fall in output in 1980. Then the long-overdue manpower savings made possible by the technical improvements of the previous decade put British Steel among the most efficient and profitable in the world. The open-hearth furnace has become a relic of the past. Most industrialized countries produce the major part of their steel by the oxygen process and the remainder from electric furnaces.

ELECTRIC FURNACE STEEL

For high-grade alloy steel cutting tools, die steels and stainless steel, the metal must be refined and melted under controlled conditions in such a way that impurities are reduced to a minimum. Where a fuel is burnt in the furnace some contamination is unavoidable; this led steel-makers to realize that electric melting was likely to be technically more effective than the methods of the Bessemer and open-hearth furnaces. The electric-arc furnace was originally intended to refine and produce alloy steels of good quality from scrap, with the addition of alloying elements. Today electric furnaces feature in the minimills, to be described later, where scrap and/or pellets of iron produced by direct reduction are converted into steel products.

An electric-arc furnace is illustrated in *Fig. 40*. The hearth can be either acid or basic lined. Acid furnaces are used mainly in steel foundries and are rarely of more than ten tonnes capacity. The much larger basic furnaces, employed for making alloy and special steels, range in size from 3 to 7 metres diameter, though there are larger furnaces in North America, including one of 11·4 metres diameter. In Britain most electric furnaces are of 7 metres diameter, with a capacity of about 170 tonnes. The furnaces usually have water-cooled linings.

The bottom of the furnace is covered with lime; scrap steel of known

Table 9. Steel tonnages per annum produced in Britain

Year	Acid Bessemer	Basic Bessemer	Acid open-hearth	Basic open-hearth	Oxygen	Electric	Others	Total
1870	215 000	—	—	—	—	—	—	215 000
1880	1 034 000	10 000	251 000	—	—	—	—	1 295 000
1890	1 613 000	402 000	1 463 000	101 000	—	—	—	3 579 000
1900	1 254 000	491 000	2 863 000	293 000	—	—	—	4 901 000
1910	1 138 000	641 000	3 016 000	1 579 000	—	—	—	6 374 000
1920	587 000	375 000	3 380 000	4 580 000	—	89 000	56 000	9 067 000
1930	279 000	—	1 805 000	5 099 000	—	76 000	66 000	7 325 000
1940	176 000	738 000	2 174 000	9 274 000	—	435 000	178 000	12 976 000
1950	248 000	846 000	1 311 000	12 981 000	—	736 000	171 000	16 293 000
1960	294 000	1 655 000	658 000	19 875 000	25 000	1 686 000	112 000	24 305 000
1970	283 000	—	162 000	13 213 000	9 102 000	5 517 000	39 000	28 316 000
1972	223 000	—	61 000	9 300 000	10 786 000	4 914 000	36 000	25 320 000
1974	139 000	—	20 000	6 168 000	10 797 000	5 271 000	30 000	22 425 000
1976	—	—	14 000	4 026 000	11 470 000	6 756 000	22 000	22 274 000
1978	—	—	—	1 764 000	11 336 000	7 200 000	11 000	20 311 000
1980	—	—	—	—	6 689 000	4 578 000	9 000	11 276 000
1982	—	—	—	—	9 036 000	4 668 000	—	13 704 000
1984	—	—	—	—	10 300 000	4 800 000	—	15 100 000
1986	—	—	—	—	10 540 000	4 040 000	120 000	14 700 000

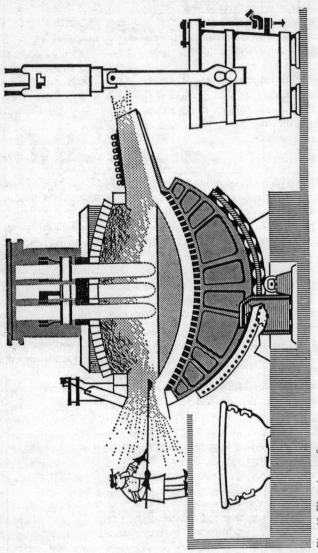

Fig. 40. Electric-arc furnace

quality is then put inside. Next the three carbon electrodes are lowered to the surface of the metal and melting begins. The slag removes much of the silicon, manganese and phosphorus from the molten scrap. Iron ore is added to remove carbon and the remainder of the phosphorus.

The furnace is tilted and the slag raked off into the ladle on the left of the diagram; it is replaced by a slag, composed of lime and fluorspar, which removes sulphur. A sample of the steel is analysed and adjustments to the composition are made by adding ferro-alloys. Finally the temperature is checked, the furnace tilted, and the metal tapped into the ladle on the right of the drawing.

STEEL PRODUCTS

When the oxygen steel-making process is complete, the molten metal is run out into large ladles, some holding as much as 300 tonnes, depending on the size and type of the plant; alloying additions are put into the ladle during pouring, to give the desired steel composition. From there, the steel is usually transferred to continuous casting machines, described later.

Most of the world's steel products come mainly within the range of 0·03 to 0·5 per cent carbon; some of their best-known uses are shown in *Fig. 43* on page 131. Liquid steel contains dissolved oxygen which can be removed by adding strong deoxidants such as aluminium and silicon. When ingots are made, three basic types are produced, depending on the mode of solidification which is governed by metal composition and the extent of the use of deoxidants. They are known as killed, rimmed and semi-killed steels.

In the production of a killed steel, virtually all the dissolved oxygen is removed and the liquid solidifies without any evolution of gas. The resulting ingot contains no porosity and is used where high strength and resistance to impact are essential, and when the steel is to be welded. Killed steel ingots shrink during solidification forming a depression or 'pipe' at the ingot top, which is discarded after rolling.

In contrast to killed steels, the manufacture of rimmed steels takes advantage of the chemical reactions which take place when a non-deoxidized steel is allowed to solidify. The dissolved oxygen combines with carbon to give carbon monoxide gas, which is released at the solidification front. The evolution of this gas gives the ingot a characteristic clean outer skin or rim. Sheet steel products requiring good surface quality, for example, car bodies and 'tin' cans, are made from rimmed steel ingots.

The third class of product is called semi-killed steel and, as the term indicates, it is only partially deoxidized prior to solidification. Some structural steel products are made from semi-killed steel.

With all these types of steel, carbon is the significant alloying element and, in general, strength increases but toughness decreases with increasing carbon content. Many modern steels for welded structures need to have carbon contents below 0·2 per cent if adequate weldability is to be ensured. In such cases greater toughness is obtained by adding small amounts of alloying elements, including manganese, nickel, chromium molybdenum, niobium and vanadium, individually or in combination, generally in the range of 0·02–1·0 per cent. The importance of carbon and the effects of alloying elements will be discussed more fully in the next two chapters.

Plate steels can be supplied hot-rolled and so may strip steels, but where a bright, smooth surface of good dimensional accuracy is required, the metal is cold-rolled, as was described on page 82. For some products, including suspension bridge cables and piano wire, the steel is made into a rod which is then drawn down to a wire of the required diameter. In this way great strength of the order of $1500N/mm^2$ is obtained.

LADLE METALLURGY

More and more stringent demands are being imposed on steel-makers by their customers: stricter and closer control of composition, cleaner steels in which inclusions are removed or made innocuous, and all this is at the lowest possible price. These requirements have led to many 'new looks' in steel manufacture; one important development involves metallurgical treatments being performed in the ladle, after the molten steel has been poured. A director of one steel plant summed up the process as 'letting the melting furnace do what it is supposed to do – make a master steel – then making the adjustments in the ladle'.

The approach to ladle metallurgy includes refining, lowering of sulphur content, control of inclusions and alloying and temperature control. There are at least seven competitive techniques involved. In one method, known as wire-feed injection, powdered reagents are encased in a metal sheath or supplied directly as a wire product. The wire is coiled on a drum stationed on a platform overlooking the ladle and is fed down into the molten steel. Some wire injectors provide calcium–silicon additions, others are used to inject alloying elements such as boron or magnesium.

In one process argon or oxygen is injected to reduce the amount of the harmful impurity sulphur. Another method adds synthetic slags to the steel in the ladle. Such after-treatments are becoming a feature in all steelworks. The trend is now to use the oxygen plant for melting and reducing the carbon content, followed by treatment in the ladle for improving the quality and for making alloying additions.

CONTINUOUS CASTING OF STEEL

A major development, in which European steel-makers have played a large part, is the production of sections direct from liquid metal. Continuous casting of steel was developed in Europe in the mid 1950s and has grown rapidly: the present world capacity is over 400 million tonnes per annum, representing about half the total tonnage of steel.

The continuous casting of steel was developed from the similar process for non-ferrous metals, described on pages 76–7, but early attempts to cast steel in this way were beset with difficulties, the most troublesome of which was the sticking of the solidified steel shell to the walls of the mould. The use of a lubricant and the introduction of a reciprocating movement of the mould were two of the many developments which helped to overcome the problems and enabled continuous casting to begin a major new era in the steel industry.

The original continuous casting machines for steel involved pouring the metal from a height of over 13 metres above ground level, with the cast section of metal descending vertically and being progressively water cooled. The operating height was then reduced by bending the section after solidification; a further reduction in height was achieved by casting the section in a curved mould and cooling on the curve instead of vertically. *Fig. 41* shows these three stages in the development of continuous casting machines. A continuous casting eight-strand machine illustrated in *Plate 12*, operating at the British Steel Corporation Lackenby Works, is highly productive, based on 'Concast', designed in Europe by a combination of British, Swiss, German and French firms. *Fig. 42* shows a machine for continuous casting of steel slabs, indicating the most important features of the equipment.

In a typical plant, steel is run into a ladle, transported to the casting bay and brought to rest above the Concast machine. A stopper rod or a slide gate is used to control the flow of steel into a tundish reservoir. Once a predetermined steel level is attained in the tundish, nozzles are opened and the molten steel feeds into the open-ended mould set in a frame with internal water cooling. At the start-up, a dummy bar,

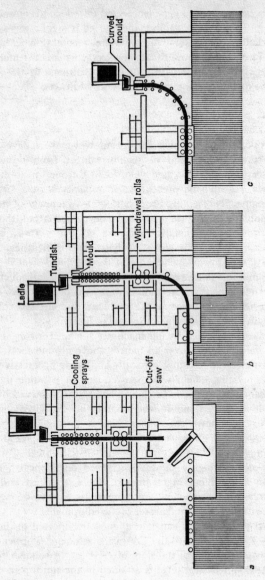

Fig. 41. Types of continuous casting machines

1. An aerial view of the Bingham Canyon copper mine – the biggest quarry on earth.

2. The West Driefontein Gold Mine, with the reduction works in the foreground.

3. One of the largest and most efficient blast furnaces in the world. This furnace at the Otia works of Nippon Steel Corporation produces over 10 000 tonnes of steel per day.

4. A Japanese electron microscope with a resolution of more than a million times. (*Courtesy Jeol (U.K.) Ltd and Xerox, Palo Alto*)

5. The Scanning Auger Microscope. (*Courtesy Materials Research Department of Defence, Government of Australia*)

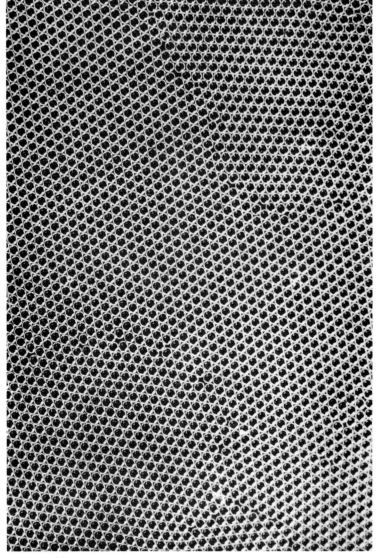

6. The Bubble-raft experiment.

7. A 70-tonne aluminium bronze propeller fitted to a V.L.C.C. of the
'Europa' class.

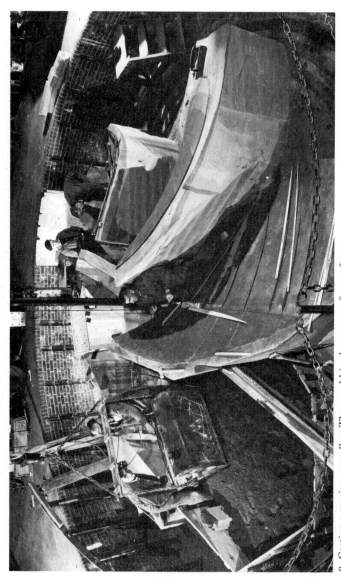

8. Casting a marine propeller. The mould in the course of manufacture.

9. The largest press of its kind in Britain. This 6000-tonne hydraulic press is capable of producing forgings up to 60kg in weight. (*Courtesy Garringtons Plant, a member of United Engineering and Forging, U.K.*)

10. Four-high mill for cold-rolling aluminium strip two metres wide, supplied to Commonwealth Aluminum Corporation, U.S.A. (*Courtesy Davy McKee, Poole, U.K.*)

11. The iron bridge over the River Severn.

12. Steel continuous casting. The British Steel Corporation's eight-strand machine at Lackenby. (*Courtesy Davy Distribution*)

13. An Ipsen vacuum heat-treating furnace with a capacity of 2·25 tonnes and maximum operating temperature of 1350°C. (*Courtesy Blandburgh Nemo Ltd*)

14. 'Eros'. A classic example of aluminium casting, which has stood in Piccadilly Circus since 1893.

15. The *Daily Telegraph* publishing and printing building, in which aluminium is a prominent feature. (*Courtesy Daily Telegraph and Alcan (U.K.) plc.*)

16. A selection of aluminium heat exchangers for automobiles. (*Courtesy Alcan (U.K.) plc.*)

17. A view of the consumable-electrode vacuum-arc furnaces for melting titanium at I.M.I.'s plant in Birmingham. The bodies of the furnaces are cooled by a molten sodium–potassium alloy. (*Courtesy I.M.I. Titanium*)

18. Rolls-Royce R.B. 211 three-shaft turbofan. (*Courtesy Rolls-Royce 1971 Ltd*)

19a. Manual metal-arc pipe welding on site. (*Courtesy E S A B Group (U.K.) Ltd*)

19b. Robotic arc welding of components. (*Courtesy E S A B Group (U.K.) Ltd*)

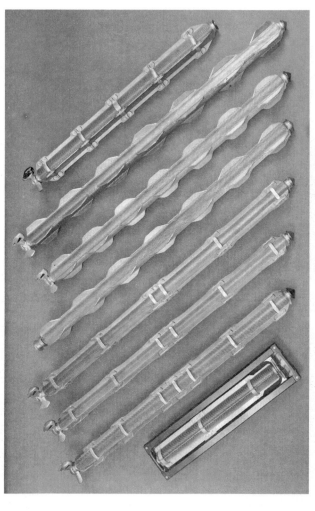

20. A selection of Magnox fuel elements manufactured at the Springfield Works of British Nuclear Fuels. (*Courtesy British Nuclear Fuels plc.*)

21. The charge-discharge machine at the AGR Hinkley Point B power station. (*Courtesy British Nuclear Fuels plc. and Central Electricity Generating Board*)

22a. Surface of cast antimony, showing dendritic structure.

22b. Pearlite. A steel with about 0.9 per cent of carbon.

22c. Martensite. The structure of a hardened steel.

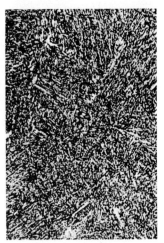

22d. The structure of a steel which has been hardened and tempered.

23a. Cuboids of antimony–tin intermetallic compound in a tin–antimony–lead bearing metal.

23b. Alloy of 60 per cent copper with 40 per cent silver, showing dendritic formation.

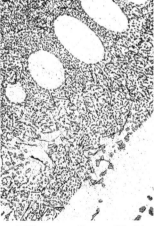

23c. Alloy containing 88·5 per cent aluminium and 11·5 per cent silicon (unmodified).

23d. The same alloy modified by the addition of 0·05 per cent sodium.

24a. Dislocations in aluminium which has been cold worked.

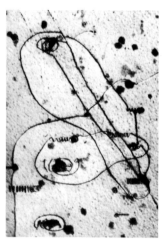

24b. Dislocation lines from particles of aluminum oxide in brass.

24c. Iron–silicon alloy showing slip-bands.

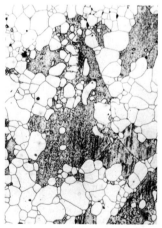

24d. The same iron–silicon alloy after annealing. The photograph shows dislocations and recrystallization.

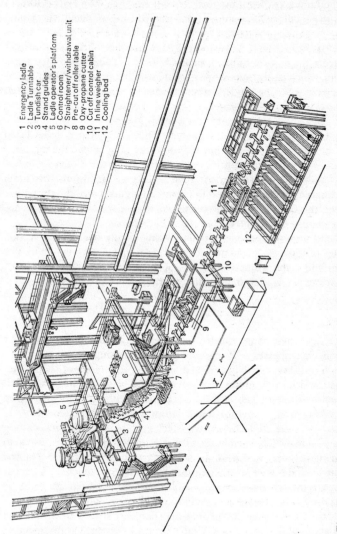

1 Emergency ladle
2 Ladle Turntable
3 Tundish car
4 Strand guides
5 Ladle operator's platform
6 Control room
7 Straightener/withdrawal unit
8 Pre-cut off roller table
9 Oxy-propane cutter
10 Cut off control cabin
11 In line weigher
12 Cooling bed

Fig. 42. A single-strand continuous casting plant for steel slab (*Courtesy Davy Distington*)

constructed of chain links, is fitted into the mould bottom and rests between the arc formed by the machine roller guides. The molten steel in the mould meets the dummy bar and freezes to it. At a certain level of steel in the mould two operations are immediately initiated: the mould unit begins to reciprocate in a vertical direction and the dummy bar is withdrawn through and down the machine. As the solidifying cast slab passes from the vertical to the horizontal plane it is supported by a series of rollers which are grouped into segments for operational and maintenance purposes. Cooling is further effected by spraying water between the rollers.

At the exit of the straightener unit a roll is lowered which separates the dummy bar from the length of cast steel. The dummy bar is either lifted to one side of the machine or hoisted on to the ramp above the machine. The strand of cast metal continues travelling along the discharge rollers to a gas torch cut-off station. At predetermined dimensions this torch will automatically cut the strand to required lengths. Once the start-up is complete, the Concast machine continues to cast sections in accordance with the volume of metal fed from the steel furnaces.

The end products of continuous casting plants are square section billets up to 150 mm × 150 mm, blooms from that size up to 400 mm × 600 mm and slabs which are up to 250 mm thick by 2500 mm wide. Other shapes are produced, including round and octagonal sections.

STEEL MINIMILLS

Minimills get their name, not because they are small, but because their operation encompasses only part of the steel-making enterprise. They are also 'mini' because they produce only a limited range of shapes and sizes within a narrow group of products. The basic design of a typical plant includes one or two electric arc furnaces either for melting scrap or for taking iron sponge or pellets from a direct reduction plant, as discussed on page 31. The molten steel goes to a continuous casting machine, from which it emerges as a continuous billet. This is cut to convenient lengths, which go to a rolling mill. The products of the minimill include wire rods, reinforcing rods for concrete and bar products, light 'I'-shaped beams and channels. High productivity is characteristic of the more profitable minimills; for example rolling mill speeds have more than doubled during the past decade and finishing rates of about 100 metres per second – ten times faster than the speed of an Olympic sprinter – are now achieved in rolling rods about 5 mm in diameter.

These plants can be located in areas where cheap fuel is available and where there is ample iron and steel scrap. An example is Texas, about 1500 miles from the big steelworks in the U.S.A., but with a great deal of manufacturing industry. It would be expensive to obtain steel from Pittsburgh and then return scrap the same long distance. The minimill provides the material and deals with the available Texan scrap. At present minimills in the U.S.A. produce a total of about 20 million tonnes per annum and it is expected that by 1990 the tonnage will increase to 25 million tonnes – about a quarter of their total steel output. Britain does not have the immense distances of America but it has some situations where the operation of minimills is logical. There is a plant at Sheerness, close to the reinforced concrete and scrap steel market near London.

Minimills have also been built in developing countries which do not have much steel scrap but have the facilities for producing iron by direct reduction. The small desert country Qatar, to the south of the Persian Gulf, illustrates what can be done with large revenues from oil, ample natural gas, technical help from some industrialized countries, cooperation with Japan to obtain export markets – and plenty of initiative. Oxide pellets received from Australia, Brazil and Sweden are reduced directly to iron in a Midrex plant. There are two electric furnaces, two four-strand continuous casters, and a rolling mill with a capacity of over 300 000 tonnes per annum.

Minimills have the advantage that they are comparatively small and are designed to make simple, locally required products. The capital cost per tonne of steel produced is only about half that for a conventional steelworks; they can be profitable in the right circumstances, though they need alert managements, capable of making rapid decisions. Such managements have been quick to adopt new methods, including electromagnetic stirring and ultra-high-power arc furnaces. One of the most effective developments of the past decade has been the use of water-cooled panels in the wall and roof construction of the electric furnace. This can make a 75 per cent reduction in the cost of furnace refractories, with a considerable lowering of production cost. Their dependence on the technology of continuous casting has been crucial to their success, and it is expected that in the next decade minimills will capture an increased proportion of the steel market. However, they can be crippled economically if the price of scrap or the cost of electric power increases drastically.

11

THE ROLE OF CARBON IN STEEL

When carbon is alloyed with iron, the hardness and strength of the metal increase; for example, a steel containing 0·4 per cent of carbon may be twice as strong as pure iron, and with about 1·0 per cent carbon, nearly three times as strong, though as the carbon content rises, the ductility is reduced. From about 1·0 to 1·5 per cent carbon the hardness increases further but the strength of the steel diminishes. Such high-carbon steels are not so much used in engineering as steels with low and medium carbon content, though tools and other instruments which are required to be very hard have carbon contents up to 1·5 per cent. Alloys of iron containing between 1·5 and 2·5 per cent carbon are rarely used. When it contains more than about 2·5 per cent carbon the metal is classed as cast iron, and characterized by good castability, moderate strength and hardness, and usually by low ductility.

For a girder or a ship's plate, a strong but ductile metal is required, capable of being fabricated cheaply and rapidly, and a steel with about 0·2 per cent carbon is employed. For a razor a hard metal is needed which can be brought to an enduring sharp edge, and the razor steel contains over 1 per cent of carbon. For guttering round the eaves of a house a cheap metal is required which can be cast to shape and which has no need to be outstandingly strong, and cast iron is therefore frequently used. So, depending on the application to which the metal is to be put, a suitable iron–carbon alloy is selected which will give the desired properties. *Fig. 43* illustrates some of the uses of steels of various carbon content.

The iron-carbon alloys are usually described according to the amount of carbon present and, though there is no precise demarcation, *Table 10* gives the usual classification.

The behaviour of carbon steels which have been slowly cooled is attributable to three causes (reference should be made to page 57).

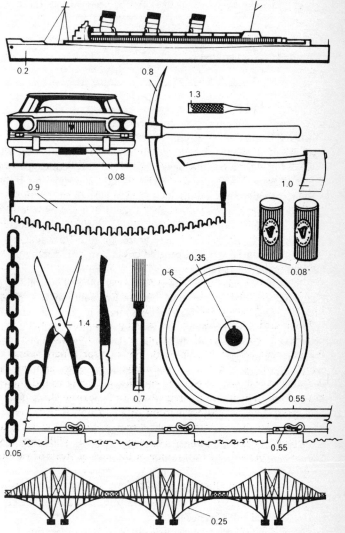

Fig. 43. Some applications of steel (Figures indicate percentage of carbon)

Table 10

Description	Approximate carbon content, per cent
Mild steel	up to 0·25
Medium-carbon steel	0·25–0·45
High-carbon steel	0·45–1·50
Cast iron	2·50–4·50

1 At room temperature the atoms of iron are arranged in the body-centred cubic lattice pattern (*Fig. 19*). When heated to 906°C, the iron atoms spontaneously rearrange into the face-centred cubic lattice pattern (*Fig. 18*). Why this takes place is not fully understood but it is a fact of great importance to a metal-using civilization, for if the change did not occur, steels would not be amenable to heat treatment.

2 At room temperature less than a hundredth of one per cent carbon can exist in solid solution in slowly cooled iron of body-centred atomic arrangement, but at high temperatures up to 1·7 per cent can be taken into solid solution in face-centred cubic iron, this maximum solubility being attained at 1145°C.

3 When molten steel containing carbon is allowed to solidify and cool slowly, the carbon does not come out of solution as such, but each carbon atom unites with three iron atoms to form a compound, iron carbide, Fe_3C. Although a compound of a metal and non-metal, it is similar in behaviour to the general class of intermetallic compounds discussed on page 52 and its presence confers increased hardness.

Carbon dissolved in iron has a pronounced effect on the temperature at which the change in the atomic lattice occurs. Thus if pure iron is gradually heated from room temperature to bright yellow heat, but not melted, the arrangement of the iron atoms changes from body-centred to face-centred cubic at 906°C. With 0·3 per cent carbon the change commences at about 730°C and is complete at about 800°C. With about 0·8 per cent the change begins and completes itself at about 730°C. Thus when carbon is present, the atomic reshuffling on heating commences at 730°C and concludes at that or some higher temperature, depending on the amount of carbon. *Fig. 44* illustrates this, and the figures given above can be checked by reference to the diagram.

The V-shaped graph is reminiscent of that representing the effect of one metal on the melting point of another, where a eutectic is formed, which has a minimum melting point. However, the behaviour of steel, which has just been described, concerns the effect of carbon on the change of atomic arrangement of solid iron. The structure of the 0·8 per cent carbon composition is called 'eutectoid' and 730°C or, to be precise, 732°C is called the eutectoid temperature. The actual eutectoid composition and temperature depend to some extent on the purity of the iron. A commercial grade of steel may show a eutectoid at 0·9 per cent and a eutectoid temperature at 700°C. For the purpose of our discussion, however, we have assumed that the iron–carbon alloy is pure.

THE STRUCTURE OF STEEL WITH 0·3 PER CENT CARBON

By referring to *Fig. 44* it is possible to picture what happens to a steel containing 0·3 per cent carbon, which has been solidified and cooled to a temperature of about 1000°C and then slowly cooled from that temperature. At 1000°C the iron atoms are arranged in the face-centred cubic pattern and the carbon therefore exists in solid solution. The upper microstructure marked *a* shows what the steel would look like at this temperature if it could be examined under the microscope. It would be seen to consist of the grains characteristic of a solid solution and might be compared with *Fig. 15* on page 50. The name given to the structure of this iron–carbon solid solution is 'austenite' after Sir William C. Roberts-Austen, a famous metallurgist of the late nineteenth century.

When this steel is cooled down to about 800°C a change takes place, indicated by a point on the upper sloping line, CE. The iron atoms begin to revert to the body-centred cubic pattern which normally holds only minute amounts of carbon in solid solution. The carbon atoms do not at once come out of solution but migrate towards areas where the iron atomic lattice is still in the face-centred cubic form, thus increasing the local carbon concentration. Finally, at about 730°C, all the remaining regions of face-centred cubic iron change over to the body-centred cubic arrangement and the carbon can no longer be contained in solid solution. This temperature is shown by the horizontal line, A E B.

By referring to illustration *b*, representing the microstructure of slowly cooled 0·3 per cent carbon steel, it will be seen that the final structure is duplex.

1 About two thirds of the structure consist of grains of iron, the white constituent in the illustration. In metallography this is identified as 'ferrite' and it is found that the presence of the grains of ferrite confers ductility on the steel.

2 About one third of the structure is a layered formation, the composition of which is discussed below.

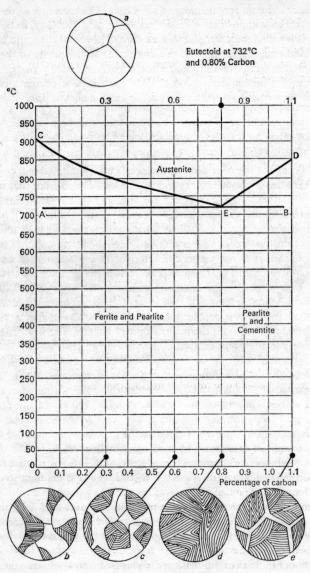

Eutectoid at 732°C
and 0.80% Carbon

Fig. 44

PEARLITE

The precipitation of carbon from the solid solution is complex. It is deposited as extremely hard iron carbide, Fe_3C. 'Cementite' is the general term used to identify this constituent, which is deposited in layers arranged alternately with layers of the surplus ferrite; this formation will be seen as the darker constituent in illustration *b*. The layered structure is called 'pearlite' because, when viewed under the microscope with oblique lighting, it produces an iridescent appearance, like that of mother-of-pearl. This formation was first seen by Professor Sorby when he was developing the use of the metallurgical microscope, in the early 1860s. A photograph of pearlite, as seen under the microscope, is shown in *Plate 22b*; it is the eutectoid which has been formed at 732°C.

With cementite alone the steel would be brittle, with ferrite alone it would be soft. Pearlite combines the good properties of both constituents, though a steel consisting of pearlite alone would be too hard for structural uses. The most widely used steel, containing 0·2 to 0·3 per cent carbon, when in the slowly cooled condition, has a structure of about one third pearlite and two thirds ferrite.

Pearlite contains about 0·8 per cent carbon and therefore a steel containing that amount of carbon is completely pearlitic when it has been slowly cooled (illustration *d*). A low-carbon steel has only a small amount of pearlite, while in a steel of over 0·8 per cent carbon the structure consists of pearlite plus surplus cementite (illustration *e*), and such steels possess great hardness and strength, but low ductility. By referring to *Fig. 44* the structure of other steels can be pictured; thus a slowly cooled steel with 0·6 per cent carbon is composed of about two thirds pearlite and one third ferrite (illustration *c*).

THE HEAT TREATMENT OF STEEL

The simple experiments described below may be enlightening. Two old-fashioned steel knitting needles, a gas ring, a bowl of water, a piece of sandpaper and a pair of pliers are needed. (Steel knitting needles are 25 cm long and are magnetic; they are not to be confused with the anodized aluminium needles which have practically replaced them and which are 30 cm long. If steel knitting needles cannot be found, a large darning needle or a bodkin will perform satisfactorily.)

1 Bend one needle slightly to feel how tough and springy it is. Now hold the needle in the flame, using the pliers, the bowl of water being close at hand. When it is bright red, dip the end of the needle as quickly as

possible into the water. The needle should still be red hot as it is being quenched. Then try to bend the quenched end; it is hard and brittle and will snap off.

2 Take the second needle and heat it until it is red hot; maintain it at this temperature for about a quarter of a minute. Then withdraw it very slowly, so that it cools gradually. If you now test this end (which you have just 'annealed') it will bend, like a piece of soft wire, and furthermore it will remain bent.

3 Heat the needle, which has been softened, to bright red heat, and quench rapidly so that it will be hard and brittle, as in Experiment 1. Clean the needle with sandpaper and hold it above the flames so that it is warmed until a straw colour develops on it; it must not reach even dull red heat. When the tint appears, take the needle away and let it cool down. If you bend the end of the needle, which you have 'tempered', you will find that the end so treated is tough and springy.

Thus three heat treatment processes have been performed on a domestic scale, processes which are being continually carried out, under rather closer control, in factories all over the world. The processes are summarized below.

Table 11

Treatment of needle	Name of process	Resulting condition of steel
Heated to red heat (850°C) and quenched	Hardening by quenching	Hard and brittle
Heated to red heat and slowly cooled	Annealing	Soft and not springy
Heated to red heat and quenched; warmed to about 250°C; cooled in air	Hardening and tempering	Tough and springy

The striking variation of properties of steel as obtained in these experiments is associated primarily with its carbon content (in this case 0·7 to 0·8 per cent), and secondly with the rate of cooling from bright red heat.

The microstructure of slowly cooled steel of the needle is similar to that shown in *Fig. 44c* and its structure consists of pearlite with some ferrite. The atomic lattice of the steel, when heated to bright red heat, changes to the face-centred cubic arrangement and the carbon is taken into solid solution. This would give the simple microstructure of austenite. From *Fig. 44* it will be seen that on *slowly heating*, the change in structure in a 0·7 per cent carbon steel starts at 710°C (see line AEB) and is complete at about 740°C (see line CED). The red-hot steel of the needle, when quenched in water as in Experiment 1, cools so rapidly that the carbon atoms have no time in which to come out of solid solution to form cementite, which is one constituent of pearlite. Because of the enforced presence of these carbon atoms the iron atoms can revert only to a distorted form of the body-centred cubic arrangement. This severe distortion is responsible for the great hardness and brittleness produced as a result of Experiment 1. Seen under the microscope this steel in the quenched condition shows a type of structure which is illustrated in *Plate 22c* and which is called 'martensite' after Adolf Martens, a nineteenth-century German metallurgist.

Few steels are used in the extremely hard martensitic condition and the re-heating operation of tempering is carried out to reduce the brittleness of the steel and yet still to retain much of the hardness, as was demonstrated in Experiment 3.

If the quenched steel is tempered at a temperature as low as 250°C its hardness is only slightly reduced but if tempering is carried out at a higher temperature the reduction of hardness is greater. Ordinary steels are not often tempered between 250° and 500°C because of a tendency in that temperature range to become embrittled, a condition known as 'blue brittleness'.

When martensite is tempered by heating it to a temperature below 700°C, particles of cementite begin to form, and in the high tempering temperature range of about 500° to 700°C the globules of cementite attain sufficient size to be seen under the microscope. The structure shown in *Plate 22d* is of a steel first hardened and then tempered at 300°C; at this temperature the cementite particles have not become visible.

Two important points are to be noted in connection with the quenching and tempering of steel:

1 To get a fully hardened structure the steel should be heated to a temperature above that represented on the line CED on *Fig. 44* and then rapidly cooled. It will be seen, therefore, that the appropriate temperature before quenching depends on the carbon content of the steel concerned.

2 Tempering does not restore the pearlitic structure. Before that can be done the steel has to be heated to a temperature above that represented by the line CED and then slowly cooled. Tempering is carried out to relieve residual stresses in the quenched steel, to produce the ultimate properties required after treatment. *Table 12* on page 147 shows the hardening and tempering temperatures of some typical alloy steels.

In the industrial treatment of steels, intermediate rates of cooling from high temperatures are also used. One treatment, known as normalizing, consists of heating a steel to red heat (above the C E D line) and cooling it in air. This gives a 'normal' ferrite–pearlite structure, similar to that shown in *Fig. 44*, though the pearlite is finer.

Although in the old days water was used as a quenching medium in the heat treatment of steel, proprietary mineral oils and sometimes gases are used today. A too severe quenching would tend to crack large blocks of steel; furthermore many modern steels contain alloying elements which dispense with the need for very rapid quenching. Heat treatment used to be done under quite primitive conditions but now it is strictly controlled and specialized, requiring very knowledgeable metallurgical supervision.

Among recent developments vacuum heat treatment is gaining popularity. The heating chamber holding the steel is evacuated, the steel is moved into it and is heated in vacuum. Then the work moves to a cooling chamber where quenching is achieved either by lowering the metal into an oil quench bath or with inert gas. Then the steel is moved to another furnace or chamber where it is tempered. The neutral conditions in the vacuum furnace prevent the occurrence of oxidation or decarburization.

There have been notable advances in vacuum heat treatment during the past five years. The impetus for this has come from expansion within the aerospace industry and the attendant requirements for high temperature treatments of titanium alloys, complex stainless steels and the nickel-based Nimonic alloys, which will be discussed on page 193. Improvements in furnace design and process control have led to larger and more efficient vacuum furnaces being built. *Plate 13* shows a vacuum furnace with a load capacity of over two tonnes and an operating temperature range from 260° to 1350°C. Its work chamber is 1·8 metres in diameter and 1·8 metres high. The workpieces to be treated are loaded on to wheel-mounted carriages which also form the bottom sealed cap of the furnace. One carriage is loaded or unloaded while the second is being treated, thus maximizing output and reducing downtime.

SURFACE HEAT TREATMENTS

Surface heat treatments are employed to improve the performance of engineering components and tools by enhancing surface hardness and resistance to wear, fatigue and, in some instances, corrosion. The techniques for ferrous metals fall into four categories:

1 High-temperature thermochemical treatments, such as carburizing.
2 Low-temperature thermochemical treatments, such as nitriding.
3 Selective surface hardening treatments, such as flame-hardening.
4 Treatments involving the deposition of ultra-hard metallic compounds, such as boronizing.

HIGH-TEMPERATURE TREATMENTS

When mild steel or certain low-carbon alloy steels are heated at temperatures above 900°C in a medium which provides a source of carbon, the carbon is absorbed at the surface of the steel by diffusion. The depth of enrichment depends on the time and temperature of the treatment, which is known as 'carburizing'. A mild steel carburized at about 925°C for eight hours should exhibit a depth of 'case' of about a millimetre and, ideally, a carbon content at the surface of about 0·9 per cent. After subsequent quenching the case-hardened steel has a hard surface layer whilst retaining the toughness of the low-carbon steel in the core.

Gears, bearings and rifle parts are a few of the multitude of components which are case-hardened. A variety of processing media is employed, in a range of furnace types. In modern industrial carburizing practice, controlled-atmosphere furnaces and salt baths predominate. In gas carburizing, components are treated in atmospheres produced either by the partial combustion of a fuel–gas/air mixture giving 'endothermic' gas or directly within the furnace by thermal decomposition of synthetic mixtures, for example, methanol with nitrogen. A common variant of gas carburizing is carbonitriding, carried out at 850–900°C with additions of ammonia to the carburizing atmosphere. This promotes the diffusion of nitrogen as well as carbon into the steel surface, enhancing the degree by which the steel may be case-hardened and permitting oil quenching rather than the more severe water quenching of mild steel items.

A carbon case can be applied by immersing the steel in a bath of molten sodium cyanide, a compound of sodium, carbon and nitrogen. Other salts, such as sodium or barium chloride and sodium carbonate, are added to lower the melting point of the mixture. Environmental considerations have prompted the development of systems in which cyanide content is reduced or eliminated. This aspect has also been partly responsible for the increasing industrial use of fluidized bed furnaces in which a heated bed of alumina particles, fluidized by appropriate gas mixtures, behaves like a liquid and furnishes an alternative medium for carburizing and carbonitriding.

LOW-TEMPERATURE TREATMENTS

Applied to previously hardened and tempered alloy steels, nitriding is a means of improving surface hardness and resistance to wear and fatigue. However, because the process temperature is only of the order of 500°C and quenching is unnecessary, components such as large gears are less susceptible to distortion as a result of the nitriding treatment. In the gas nitriding process, workpieces are heated in ammonia which partially dissociates into hydrogen and nitrogen. The nitrogen diffuses into steels containing aluminium, chromium or vanadium to produce alloy nitrides with a high surface hardness. Recent years have witnessed the wide use of plasma nitriding under vacuum conditions where regulation of electric parameters offers precise control of the resultant nitrided layer structure. As a low-temperature process, nitriding often involves long cycles and is correspondingly more expensive than carburizing-type case-hardening treatments.

A short-cycle derivative of nitriding, 'ferritic nitrocarburizing', is being employed to upgrade the performance of ferrous parts, particularly low-carbon non-alloy steels. Carried out at temperatures around 570°C, typically for two hours, the process enriches the surface with both nitrogen and carbon. It produces a thin compound layer of iron nitride that enhances wear and corrosion resistance, supported by a nitrogen diffusion zone which, if parts are quenched after treatment, improves fatigue strength. Nitrocarburizing can be conducted in a variety of processing media including salt (e.g. 'Tufftride') and ammonia-rich controlled atmospheres. Additional treatment can be applied to confer corrosion resistance rivalling that of hard chromium plating.

SELECTIVE TREATMENTS

The case-hardening treatments described above can be defined as 'thermochemical' processes, involving a combination of thermal treatment and chemical reaction. The third important category, relying on thermal input only, is represented by the induction and flame-hardening processes used in industry to harden selected areas of engineering components such as gears, crankshafts and camshafts. The treatments involve heating the surface of medium carbon or alloy steel components to a temperature at which the carbon enters into solid solution in the iron. This is known as austenitizing and can be understood by reference to *Fig. 44* on page 134; it will be seen that the temperature at which austenite is formed depends on the carbon content. Then the hot steel is quenched in oil or water to change the structure near the surface to a martensitic condition, which was discussed on page 137, thus providing a very hard surface to the component.

One of the latest surface-treatment techniques is laser hardening which has the big advantage that selected areas can be hardened without treating the whole part. Because the heat is concentrated intensely at the surface, little distortion takes place – which means that minimal post-treatment finishing is required. Ion implantation is another new surface-hardening technique; nitrogen ions (electrically charged atoms) are bombarded at metal components. After treatment for between two and ten hours a sub-surface layer is created to a depth of 0·15 microns. No distortion occurs and the life of the component is increased substantially.

DEPOSITION OF ULTRA-HARD COMPOUNDS

The steels used for tools and dies have been a particular focus of new surface treatments which impart ultra-hard surface compounds such as borides, nitrides and carbides. These include the Toyota (TD) process, developed in Japan, to produce compound layers of carbides of vanadium, niobium and chromium on the steel, which is treated in molten salts at a temperature of up to 1050°C. The physical vapour deposition process (PVD) is being exploited to improve the performance of cutting tools by imparting a very thin (about three thousandths of a millimetre) coating of gold-coloured titanium nitride, which is extremely hard.

Modern heat treatment technology permits industrial processes to be conducted on a sound scientific basis, often with sophisticated controls incorporating computerized systems. It remains a fertile field for innovation, driven by the need to enhance product quality, conserve energy, contain costs and comply with new manufacturing philosophies. The science of surface hardening has come a long way since a mixture of viper flesh and mummy dust was used by the ancient Egyptians for carburizing iron. As with many other aspects of metallurgy, art is being transformed into science and the current trends of heat treatment are towards higher throughputs, lower energy use and reduced environmental pollution, while the industry is establishing codes of practice with emphasis on design, training and safety.

12

CAST IRON AND ALLOY STEELS

Pig iron from the blast furnace is generally converted to steel, but a related ferrous alloy, cast iron, is made by remelting the metal in cupolas; these look like small blast furnaces but they melt the pig iron and do not smelt it from ore. Typical foundries produce over 500 tonnes per week, consisting of about 250 000 individual castings. Such a foundry would have two cupolas, each capable of melting 20 tonnes of cast iron per hour, with one cupola in use and the other being patched and prepared for the next day's operation.

Ten years ago less than 5 per cent of cast iron in Britain was produced using oxygen injected in the air blast but nowadays about 20 per cent is made in this way. Many cupolas operate what is known as divided blast, the air being injected through two rows of tuyères about 900 mm apart, thus increasing the melting rate and reducing coke consumption. Another development has been the melting of cast iron in electric furnaces. Cupola charges now contain only 10–20 per cent pig iron and about the same amount of steel scrap. The rest is cast iron scrap, which yields a cheaper and slightly stronger production than remelted pig iron. Modern 'press button' chemical analysis has made it possible to control and adjust the metal composition before casting.

The manufacture of all sorts of engineering and structural components in cast iron represents an important part of the metal-working field. The great fluidity of cast iron and its low shrinkage on solidification make possible close tolerances and considerable freedom in design. Cast irons have versatility in properties, including hardness, good machinability, excellent compressive strength, rigidity and resistance to wear. Cast iron is the cheapest metal, with a vast range of end products, though the tonnages required have fluctuated. Ten years ago about 300 000 tonnes of cast iron were required for ingot moulds in Britain but, thanks to the rapid development of continuous casting, less than 100 000 tonnes of cast iron ingot moulds were made in 1987. Builders' hardware has changed little during the past decade, about 80 000 tonnes of cast iron being required. One of the largest uses of cast iron is the field of pressure pipes and fittings, nearly 200 000 tonnes per annum being required in the U.K., mostly as spheroidal graphite iron, which will be discussed later. The use of cast iron in the automobile industry depends partly on the current state of the economy and on the ever-changing competition between ferrous metals and light alloys. One thing is certain, the cast iron industry has been affected by, and in the long run has benefited from, the emphasis now placed by all consumers on quality control and accuracy.

It is pleasing to report that there has been a resurgence in the use of cast iron in the streets of towns and cities. While the cast splendours created by our Victorian forefathers may never be repeated, many towns are installing well-designed castings with an old-fashioned look as bollards, lamp standards, drinking fountains, bandstands, litter bins and a Chinese gateway in Gerrard Street, Westminster.

Several varieties of cast iron can be produced, by selection of different pig irons, by variations of the melting conditions and by special alloying additions, but in general the two main classifications are into white cast irons and grey cast irons. Both contain 2·4–4·0 per cent of carbon but the difference lies in the condition in which the major portion of the carbon exists in the structure of the metal. In white cast iron all the carbon is present as cementite, and the fracture of such an iron is white. In grey cast iron most of the carbon is present as flakes of graphite, and there is usually a remainder which is in the form of pearlite; the fracture of this type is grey.

Thomas Turner, one of the great metallurgists of the nineteenth century, who later became the first Professor of Metallurgy in the University of Birmingham, published a historic article entitled 'Influence of silicon on the properties of cast iron' in the *Journal* of the Chemical Society in 1885. His research was a landmark in what we described in the first chapter as the art of metal-working growing into the science of metallurgy. Professor Turner proved that the amount of silicon present determines the condition of cast iron. With only 1 per cent silicon the iron tends to be white, while with about 3 per cent silicon, even rapidly cooled irons are grey. Later it was found that the presence of other alloying elements also had an effect on the structure of cast iron; for example, chromium tends to produce a white cast iron and nickel to produce a grey one. Alloying elements are added to some cast irons to improve their resistance to wear and corrosion.

Since cementite is intensely hard, white cast iron is hard and durable, though very brittle. Grey cast iron is softer, readily machinable, less brittle, and suitable for sliding surfaces because graphite is soft and is a good lubricant. The rate of cooling to some extent determines whether the iron is white or grey; the more rapid the cooling the greater the tendency to form a white iron. Use is made of this fact in producing large cast-iron rolls for rolling mills, which have a grey centre but a chilled, and therefore a hard, white iron surface.

White cast iron is used to a much lesser extent than grey, although its hardness and wear-resistance make it suitable for components of mining and quarrying machinery. However, it is an important stage in the manufacture of malleable cast iron, which is discussed later.

SPHEROIDAL GRAPHITE CAST IRON

Since 1948 spheroidal graphite cast iron, better known as S.G. iron or nodular iron, and in the U.S.A. as ductile iron, has been developed as a constructional material which can compete with steel castings or forgings for many stressed components.

In the search for improved properties it was discovered that the addition of about 0·04 per cent of magnesium or cerium to molten cast iron would cause the graphite to form into small nodules when the metal solidified. This allowed the treated cast iron to be stronger, tougher and more shock-resistant than ordinary cast iron in which the flakes have a comparatively large surface area. This spheroidal graphite iron is often melted in electric furnaces because they provide greater purity than can be obtained when melting in cupolas. More than half of all the spheroidal graphite iron produced is centrifugally cast, to make pipes of diameters ranging from 50 mm up to 1000 mm, used for carrying water and gas at both high and low pressures.

One of the most interesting markets is the car industry, which is always on the look-out for price savings without loss of quality. European manufacturers are now using S.G. iron for crankshafts; another development is its use for the calipers of disc brakes, where high quality performance and safety are of paramount importance.

MALLEABLE CAST IRON

White cast iron is the basis for the production of malleable iron; it is heat treated in such a way that the white cementite is broken down into ferrite and graphite of a structure which tends to be nodular in form and gives a

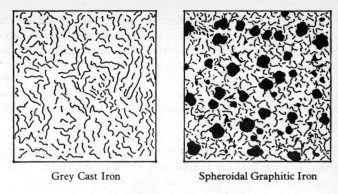

Grey Cast Iron   Spheroidal Graphitic Iron

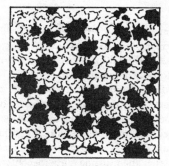

Blackheart Malleable Iron

Fig. 45 Microstructure of three forms of cast iron

black fracture. Such iron is then known as blackheart malleable iron. *Fig. 45* shows the microstructure of the three types of cast iron. In the grey cast iron the flakes of graphite can be seen. The spheroidal graphite is shown as dark, rounded nodules, while the form of the graphite in malleable iron consists of nodules of a fuzzier type than those of S.G. iron.

Pearlitic malleable iron was developed during the Second World War as a good substitute material for steels. It can be produced by increasing the amount of manganese in white cast iron and heat treating it. Different grades are produced either by air quenching or oil quenching followed by tempering, and the matrix structure will vary from fine pearlite to tempered martensite. Most of its present uses are for parts which were originally produced in steel.

Malleable iron is an old industry, the process having been discovered two hundred years ago. Perhaps it has the benefits as well as the disadvantages of a process which has been modernized and expanded during the last forty years. The industry makes a product which is inexpensive and suited to the needs of modern production requirements because malleable iron is strong, tough and can be machined rapidly. S.G. iron is a strong competitor with malleable iron and is tending to replace it in circumstances where in the past their uses overlapped. However malleable iron components are usually less than 60 mm thick, because it would be difficult to produce a very thick white cast iron, which is the starting point for producing malleable iron. On the other hand, S.G. iron is not so suitable for use in thin sections.

ALLOY STEELS

A thin piece of steel such as a small tool or needle can be quenched and will give uniform properties throughout its section, but if a massive block of steel is heated and quenched the interior will cool fairly slowly, despite the rapidity of cooling on the outside. Thus there is a gradual decrease of hardness from the outside to the centre of the quenched piece. Furthermore the unequal dimensional behaviour of inside and outside causes internal stresses to be set up, so that a heavy block of high-carbon steel may crack if it is drastically quenched. Such a state of affairs places limitations on the use of carbon steels, but these difficulties have been overcome, and other advantages conferred, by the development of alloy steels, in which uniform hardness can be produced throughout the whole of a block by a mild quench.

Alloy steels are divided into two types: low alloy steels with under 10 per cent of added elements and high alloy steels with over 10 per cent (usually between 15 and 30 per cent) of added elements. During the past hundred years, alloy steels have entered more and more into the manufacture of highly stressed components or parts which have to work in corrosive conditions, from armour plate, bridges and bicycles, right through the alphabet at least as far as washing machines and yachts. Housewives turn up their noses at ordinary steel knives – only stainless alloy steel cutlery is acceptable.

Alloy steels are used for the following reasons:

1 To enable the effect of heat treatment to appear uniformly in large masses of steel without the outer skin behaving differently from the inside, as would be the case if bulky sections of 'straight' carbon steel were treated.

2 To obtain a given combination of mechanical properties by less drastic

Table 12

Type of Steel	Approximate Composition	Temperature °C	
		Hardening	Tempering
Die Steel	0·4% Carbon 5% Chromium 1% Vanadium 1·5% Molybdenum	1000	600–650
Tool Steel	1·5% Carbon 13% Chromium	980	175–420
Air Hardening Tool Steel	1% Carbon 5% Chromium 0·5% Vanadium 1% Molybdenum	970	380–400
Nitriding Steel	0·3% Carbon 0·35% Silicon 1·6% Chromium 1·1% Aluminium	900	500–720

heat treatment than would be necessary in plain carbon steel, thus minimizing internal stresses.

3 To impart to the metal special qualities such as great strength, hardness, resistance to wear, springiness or resistance to corrosion.

In order to get the full value of improvements conferred by the alloying elements, these steels are generally used in the heat-treated condition. Carbon is an essential constituent of the steels which are heat-treated, for without carbon they could not attain their useful mechanical properties; the carbon makes hardening and tempering possible, the alloying elements *modify* the effect of the treatment. One of the earliest and most famous alloy steels was that containing nickel, discussed on page 191. Nickel steels were later improved by the addition of chromium. Such steels proved to be somewhat temperamental in tempering; they developed a mysterious embrittlement, but a small addition of molybdenum was found to cure this 'temper brittleness'.

Various combinations of strength, hardness, springiness and toughness may be achieved by the selection of times and temperatures of heat treatment for a particular alloy steel composition. The field is so vast that only a brief selection could be given in the table above.

A landmark in alloy steel history was the discovery of manganese steel by Sir Robert Hadfield in 1882. Such a steel, containing about 1 per cent carbon and 13 per cent manganese, can be brought to a high degree of toughness by heating it to 1000°C and quenching in water. In this state the structure of the steel is austenitic and it is only moderately hard, but any attempt to cut or abrade the surface results in the local formation of hard martensite, so that the steel cannot be machined by ordinary cutting tools. Manganese steel is therefore used for purposes which require intensely hard metal, such as parts of rock-breaking machinery and railway crossings. It is used for the bars of prison cells, because the steel makes it almost impossible to escape by the well-known method of using a file.

STAINLESS STEEL

In 1913 Harry Brearley of Sheffield was experimenting with alloy steels for gun barrels, and among the samples which he threw aside as being unsuitable was one containing about 14 per cent chromium. Some months later he noticed that most of the steels had rusted, but the chromium steel remained bright. This led to the development of stainless steels, which possess a very high resistance to corrosion due to a naturally occurring chromium-rich oxide film, invisible and sub-microscopically thin; if the surface film is damaged by scratching, it is self-healing and protective.

The stainless steels, which now often contain other elements besides chromium, are among the most corrosion-resistant alloys available and their use has doubled during the past decade. Now about 5 million tonnes are made per annum. The Scandinavian countries have done a great deal to make stainless steel articles elegant and popular. About 7 per cent of the total Swedish steel production is stainless, compared with about 1 per cent for Britain and 2 per cent for Japan. There are three important categories of stainless steel:

1 The Martensitic group contains between 12 and 16 per cent chromium. With carbon at less than 0·15 per cent this type of steel is used for cutlery; when sharper cutting properties are required the carbon is increased to about 0·7 per cent.

2 The Ferritic group contains between 12 and 16 per cent chromium but the carbon content must be no more than 0·1 per cent. These steels cannot be hardened by heat treatment but they can be strengthened by cold working. They are used for car trim and commercial catering equipment.

3 The Austenitic group, which is produced in the greatest quantity, contains 15 to 25 per cent chromium, with 7 to 12 per cent nickel. Best

known is the steel with 18 per cent chromium and 8 per cent nickel, used for kitchen sinks and beer barrels. Molybdenum, up to 3 per cent, is added for still greater corrosion resistance, required for example in superheater tubing in power stations and the cladding of buildings in polluted atmospheres. Although they are called steels, the materials in this group perform better if they contain no carbon, and special techniques are used to reduce the carbon content as low as possible.

By adding elements such as 3 per cent copper or aluminium with about 1 per cent titanium to steels containing 17 per cent chromium and 4 to 8 per cent nickel, heat treatment involving precipitation hardening and/or refrigeration at $-70°C$ can be applied, and strengths of up to 1500 newtons per sq. mm can be obtained. Such steels are used for aircraft undercarriages and honeycomb stiffened structures for the walls and skinning of missiles and space craft.

There is a range of at least ten austenitic stainless steels, graded according to their ability to resist severe corrosion. The first steel in the list contains 17 per cent chromium, 12 per cent nickel and 2·25 per cent molybdenum, while the one with maximum resistance to corrosion contains 22 per cent chromium, 5 per cent nickel and 3·6 per cent molybdenum, plus a small but important content of nitrogen. Most of the steels contain only a small amount of carbon, of the order of 0·03 per cent.

A great deal of new equipment has been designed for the fabrication of stainless steel, to meet the increasing demands. The largest consumers are the domestic equipment and automotive industries. In Britain the manufacture of sinks is the largest single use. Stainless steels are widely used in automobiles for external bright trim. Stainless steel exhaust pipes are, rightly, becoming popular and it has been ruefully remarked that they are so durable that they are likely to outlast the rest of the vehicle. Their use in the nuclear energy industry will be discussed in Chapter 23. Some other recent developments include telegraph poles and call boxes. The brewery, dairy and chemical industries are now using considerable tonnages and stainless steel razor-blades have displaced ordinary steel.

The success of the modern razor-blade is due to the presence of a thin layer of the plastic P.T.F.E., which has to be baked on after the blade is sharpened. Conventional razor-blade steel softened when the plastic was being baked on, and was replaced by a martensitic stainless steel containing 13 per cent chromium, 0·67 per cent carbon, 0·6 per cent manganese, and 0·4 per cent silicon. Precise metallurgical control of the heat treatment is necessary to ensure even distribution of the martensite particles. Stringent quality control is also required in the manufacturing operations, including pressing, clipping, the deep freeze which follows heat treatment, and finally packing.

When manufacture began to expand world-wide twenty years ago, the stainless steel was produced by melting steel, partly scrap, in electric arc furnaces, adding the required alloying elements as ferro-alloys and casting the steel into ingots. Now, thanks to the great developments in continuous casting, manufacturers are able to supply stainless steel in a condition much nearer to the finished product. Electric arc furnaces have the disadvantage that the high temperature causes some loss of chromium but this problem has been overcome by what is known as argon–oxygen decarburization. First the metal is rapidly melted in an electric-arc furnace, but not at such a high temperature that chromium loss occurs. Then an argon–oxygen mixture is injected into the molten metal which refines it and avoids loss of chromium.

Stainless steels have been used extensively for architecture and the trend is growing. New buildings have included large quantities of stainless steel for curtain walling, decorative panels, and door and window frames. Considerable amounts are used for components in power stations, particularly in the field of nuclear power production. As science advances from guided missiles to supersonic aircraft and flights into outer space, further use will be made of stainless steels because they provide such an excellent combination of resistance to heat and corrosion, and maintenance of high strength at elevated temperatures.

HIGH-SPEED STEELS

One other type of alloy steel, which is important in the history of metallurgy, and indeed of civilization, is the high-speed steel used for cutting metals on lathes and other machine tools. Before A.D. 1900 a cutting speed of 10 metres per minute was considered good for ordinary carbon steel tools; great astonishment was caused at the Paris Exhibition in that year when the Bethlehem Steel Corporation of America exhibited cutting-tool steels, containing tungsten and chromium, that would cut for hours at 50 metres per minute and continue to cut for some time even when the speed was so much increased that the tip of the tool became red-hot. The new type of steel was developed by Fred W. Taylor and Maunsel White, and later they recommended the use of a steel containing about 0·7 per cent carbon, 18 per cent tungsten, and 4 to 6 per cent chromium. This remarkable stability at high temperatures was largely due to the presence of tungsten carbide, and some chromium carbide.

In addition to iron, carbon, tungsten, and chromium, high-speed steels may contain vanadium and cobalt. Since the discovery of tungsten steel, other tool materials have been developed, for example, molybdenum steels. But one of the best metallic cutting materials turned out to be tungsten carbide bonded with about 10 per cent cobalt. In fact this material is all alloying elements and contains no iron.

NEW DEVELOPMENTS IN ALLOY STEELS

The consumption of alloy steel is going up steadily and efforts have been made to economize in the scarce and expensive elements which are essential constituents of these steels. Sometimes the addition of minute amounts of new elements gives an effect equivalent to that obtained by much larger amounts of other alloying metals. Recent discoveries have led to the addition of small amounts of boron to increase the hardenability of steels of low alloy content. One steel containing 0·15 per cent carbon, 0·40 per cent molybdenum and 0·003 per cent of boron has a high strength and good weldability.

The high-speed steels which are used for cutting tools illustrate the conservation of alloying elements, which became necessary during the 1939 war and which, once again, has become a matter of great concern, owing to the shortages and high prices of many elements used in the manufacture of alloy steels. It was found that high-speed steels could be made with less alloying content than was formerly considered essential. Steels containing 3 per cent each of tungsten, molybdenum, and vanadium were introduced in Germany; other steels, developed in Britain and the U.S.A., contained 0·8 per cent carbon, 6 per cent tungsten, and 4 per cent molybdenum. These alloying contents are to be compared with 14 to 22 per cent of tungsten which was the normal pre-war range of composition for a high-speed tool steel. Raising the vanadium content to as much as 4 to 5 per cent has proved successful in some high-speed steels.

The high-strength low-alloy (H.S.L.A.) steels contain about 0·07 per cent of vanadium, sometimes with smaller amounts of niobium and titanium, the carbides of these elements being dispersed through the steel, providing a considerable increase in strength. The H.S.L.A. steels are used for bridges, high-rise buildings, aircraft hangars, oil and gas pipe lines and earth-moving equipment.

Another material is known as dual-phase steel; a typical composition is

0·07 per cent carbon, 1·6 per cent manganese, and 0·85 per cent silicon. Their structure is of an alloyed ferrite matrix, dispersion hardened with between 10 to 20 per cent martensite. These steels are playing an important part in the making of car body panels, wheels and chassis members, where high strength and rigidity must be combined with ductility to enable press-forming operations to be carried out efficiently.

SUPER-ALLOY STEELS

Stringent control of manufacture and the demand for optimum mechanical properties have led to the development of super alloy steels which are completely free from non-metallic inclusions. Such inclusions may originate from refractories with which the steel is in contact in melting and casting; they are potential sources of failure when severe stresses are encountered, as in modern aircraft engines. The consumable-arc melting technique, used for molybdenum and titanium (illustrated in *Plate 17*), is now being adopted for melting super-alloy steels. Such furnaces can melt up to 8000 kg.

Vacuum melting is being used increasingly for improving the quality of steels. This technique helps to prevent the formation of non-metallic inclusions and it eliminates deleterious gases. In the older and still popular process the steel is melted in an evacuated vessel. The process is now being extended to larger tonnages by enclosing the ladle into which the steel is cast within a vacuum chamber.

Another method of extracting the last traces of inclusions is by electro-slag melting. This was originally developed in the U.S.A. mainly for welding thick sections of steel, but later the U.S.S.R. extended the idea to refine steels by remelting steel continuously under a flux, the energy being supplied by an electric current across the slag pool.

13

ALUMINIUM

Aluminium, known as aluminum in America and many other countries, is the second most important metal. During the past hundred years its production has increased a million times. Even since 1939 output has expanded substantially; in that year the world production was about 750 000 tonnes and by 1943 it had increased to two million tonnes. After the war production fell temporarily but then started to rise; now the world's output is about sixteen million tonnes per annum, to which must be added over five million tonnes obtained from recycled scrap. The industry has done a great deal to develop new uses for the metal, often with bold and imaginative pioneering. The use of aluminium in aircraft, shipping, automobiles and building was made possible by dedicated research and development in the design of components and new methods of fabrication and joining. Several properties account for the importance of aluminium.

1 It is a light metal with a density of 2·7 – only about a third of that of steel or brass;
2 Aluminium has a higher resistance to atmospheric corrosion than many other metals, owing to the protection conferred by the thin but tenacious film of aluminium oxide which forms on its surface;
3 Aluminium is a good conductor of electricity. For the same cross-section area of wire the metal has about two thirds the electrical conductivity of copper. But aluminium is a better conductor on a weight-for-weight basis, because of its lower density;
4 Aluminium conducts heat well;
5 Aluminium forms high-strength alloys in conjunction with other elements. Some of these alloys are only one third the weight of mild steel per unit volume and of equal strength;
6 A large number of useful aluminium alloys can be cast, rolled, extruded, drawn, pressed, riveted, machined and welded without much difficulty.

ALUMINIUM OXIDE AS A PROTECTOR

The graceful figure of Eros, illustrated in *Plate 14*, was sculpted by Sir Alfred Gilbert and cast in a London foundry. After over eighty years of exposure to the Piccadilly Circus atmosphere, Eros shows little sign of corrosion.* Other examples of the corrosion resistance of aluminium are in food manufacture and storage, in petroleum refining and in the manufacture of nitric acid and explosives where it has replaced fragile earthenware containers.

However, aluminium can be a very reactive element and the metallurgist knows that its successful resistance to corrosion depends on the completeness with which the protective layer of aluminium oxide prevents this underlying, lurking activity from coming into play. As soon as aluminium is cut and exposed to the air, a hydrated aluminium oxide forms on the fresh surface. It seals the surface of the metal so that further oxidation is prevented. Other metals are not so fortunate; corrosion products of iron are soft and crumbly and do not prevent progressive attack.

The natural film of aluminium oxide is only a few thousandths of a millimetre thick. It can be thickened up to between 0·003 mm and 0·025 mm by an electrolytic process known as anodizing. The aluminium articles are suspended in vats similar to those used in electroplating, but usually containing sulphuric acid solution; the components to be anodized are at the positive, anode, end of the vat. The action of the electric current releases oxygen from the solution and a tenacious coating of aluminium oxide forms on the surface of the metal. Depending on the operating conditions, anodizing can be adjusted to produce either a film of great hardness or one that is very resistant to corrosion. The anodic film also possesses the ability to absorb dyes, thus enabling the metal to be tinted with attractive and enduring colours. For special purposes a hard anodic film, up to a fifth of a millimetre thick, can be produced.

ALUMINIUM AS ELECTRIC CONDUCTOR

Aluminium was used as an overhead conductor over fifty years ago, first in Switzerland and then in North America. Later its use became more general when it was found possible to reinforce the aluminium conductor by a central core of galvanized steel wire. The resulting increase in

---

* Although Eros was restored in 1986 and 1988, no corrosive damage to the metal was discovered. The crack in one leg and the loose joints were inflicted by human interference alone.

strength of the cable made long spans possible without too great a sag. This 'aluminium conductor, steel-reinforced' (A.C.S.R.) is used in practically the whole of the British National Grid. Severn unjointed lengths of A.C.S.R. conductor cables, each 4 kilometres long and weighing 23 tonnes, carry an important link in the Central Electricity Generating Board's super-grid system over the rivers Severn and Wye. The crossing of the two rivers at their confluence involved three long spans, one of them nearly two kilometres in length – the longest in Britain. Each conductor consists of 78 aluminium wires of 2·8 mm diameter, arranged in two layers around a core of 91 galvanized high-tensile steel wires of the same diameter. The seven 4-kilometre lengths involved stranding 4½ million metres of steel and aluminium wire. The main span, 1600 metres long, has a sag of 60 metres, giving a clearance of 40 metres above the water. The wires in the outer layer are given a trapezoidal shape to provide a smooth surface and to minimize wind resistance, in order to avoid developing harmonic vibrations.

SOME USES OF PURE ALUMINIUM

Pure aluminium foil, formerly called 'silver paper', is produced by hot-rolling aluminium ingot till it has been reduced to between 5 and 10 mm thick, after which the metal is cold-rolled, with intermediate and final anneals. The end products are coils of aluminium foil – 0·008 mm thick for wrapping chocolates and cigarettes; domestic foil 0·015 mm thick; 0·04 mm thick for milk-bottle tops, cream and yogurt lids; and 0·1 mm thick for food containers.

Convenience foods are mass-packaged with the aid of foil; the mix is put into individual dishes which travel on a conveyor in the cooking oven; the product, when prepared, is transported and sold in the same foil container. This is a thriving industry; in the U.K. over a thousand million foil food containers were manufactured in 1986, requiring about 11 000 tonnes of aluminium. Meat and poultry for roasting can be wrapped like a parcel in foil, after being covered with fat; no basting is necessary, and the foil wrapping retains the juices and reduces the escape of cooking odours.

The industry is breaking into new fields with considerable enterprise. Peel-off lidding for metal and plastic containers of pharmaceutical products and convenience foods is a boon. Do-it-yourself car repairs have been made easier with foil kits; exhaust pipes are repaired with foil bandage impregnated with adhesive. In the pop music scene an embossed and laminated aluminium foil makes attractive record sleeves. The same type of material is being used in bars to provide patterns and colours which change to the beat of the music.

Aluminium is a good reflector of heat; storage containers for petrol, milk, and other liquids are coated with crinkled aluminium foil which reflects away the sun's heat instead of absorbing it and so the liquid inside is prevented from becoming unduly heated. Also, for the same reason, aluminium is used in building to maintain constant temperatures in hot and cold weather; the efficiency of aluminium foil for insulation purposes depends upon its being associated with a still air space.

Polished or electro-brightened high-purity aluminium is one of the best materials for reflecting light, far better than ordinary mirror-glass and in some respects even more suitable than silver. Aluminium-coated reflectors have two advantages of special interest to astronomers; first, when the front of an astronomical telescope mirror is coated with aluminium it does not tarnish so rapidly as silver; secondly, aluminium reflects ultra-violet light better than silver.

The use of an aluminium mirror coat is now standard practice for all large reflecting telescopes, including the 200-inch Hale telescope at Palomar mountain. This famous telescope is situated in a dry climatic zone, in California; it is re-aluminized about once every five years and is carefully washed every few months. Mirrors in wetter climates deteriorate much more rapidly; for example, when the Isaac Newton telescope was located at Herstmonceux, near the coast of Sussex, the $2\frac{1}{2}$ metre mirror lost reflectivity within a few months, due to salt in the air. Therefore it had to be re-aluminized every spring, summer and early winter: the aluminium was stripped off, the mirror cleaned, polished and dried, with a final holding in vacuum; the fresh coating of aluminium was then applied. During the early 1980s this famous telescope was transferred to Las Palmas Observatory, high on a mountain in the Canaries, where conditions are more suitable. Aluminized mirrors can be protected by evaporating a very thin layer of transparent fused silica on their surfaces. However such coatings transmit only a limited range of colours. New silica-based materials are now being developed which do not suffer this disadvantage and they are being used for mirrors up to 50 cm in diameter.

Very many buildings constructed since the war have employed aluminium curtain wall structures, the metal being either commercially pure or with only a small amount of alloying elements. *Plate 15* shows the new printing and publishing headquarters of the *Daily Telegraph* in the Isle of Dogs. In keeping with the London Dockland surroundings the building was designed to resemble a ship. The upper parts were of aluminium with about 0·5 per cent of each of magnesium and manganese. Profiled lengths 7·5 metres long were produced from coils which had been etched in caustic soda, treated with chromate and then coated with a polyvinylidene fluoride paint system.

ALUMINIUM CASTING ALLOYS

There is a wide range of alloys available for casting, and a variety of properties which determine the alloy selected for a particular use. Silicon, copper and magnesium are used as the main alloying elements, either alone or in combination. Often small amounts of other elements are added, including manganese, zinc, titanium and nickel. Most of the alloys have a permitted allowance of about one per cent of iron. All of them can be cast in sand moulds and nearly all can be gravity diecast; a limited number are suitable for pressure diecasting.

The major part of the 80 000 tonnes of aluminium alloys cast in Britain each year is selected from the 'LM' (light metal) alloys specified by the British Standards Institution, the list being revised every few years to take advantage of new developments. The available alloys in the LM range total over twenty, but of these, eight alloys account for about 85 per cent of the total: *Table 13* shows these alloys. There are other aluminium casting alloys, which are patented compositions including those developed for uses in aircraft.

The alloy LM6, containing 10 to 13 per cent of silicon, shows an interesting metallurgical phenomenon known as 'modification'. As normally cast in a sand mould the alloy has a coarse structure, shown in *Plate 23c*; this is accompanied by low strength and weakness under shock. In 1920 Dr Aladar Pacz discovered that a small addition of the compound sodium–potassium fluoride brought about a change in structure. It was found that other compounds of sodium, or about 0·05 per cent of the metal itself, added to molten aluminium–silicon alloy, modify the structure obtained on solidification (*Plate 23d*). The alloy treated under these conditions is stronger and tougher than the unmodified alloy; it has an excellent combination of good strength, ductility, resistance to corrosion and good castability. A foundry foreman we once knew was heard to remark, 'it runs like milk'.

An alloy known as 380 in the U.S.A. and LM24 in the U.K. contains between 7·5 and 9·5 per cent silicon and from 3 to 4 per cent of copper. It can be produced from scrap and is therefore comparatively inexpensive. It is more suitable for rapid machining than the 'straight' alloy of aluminium with silicon and this property is particularly

Table 13. Some important aluminium casting alloys

Alloy Specification BS1490	Main alloying constituents Percentage (single figures are maxima)						Approximate percentage of British consumption 1987 (%)
	Copper	Silicon	Iron	Nickel	Magnesium	Zinc	
LM2	0·7–2·5	9·0–11·5	1·0	0·5	0·3	2·0	8
LM4	2–4	4–6	0·8	0·3	0·15	0·5	18
LM6	0·1	10–13	0·6	0·1	0·1	0·1	14
LM9	0·1	10–13	0·6	0·1	0·2–0·6	0·1	2
LM13*	0·7–1·5	10–13	1·0	1·5	0·8–1·5	0·5	4
LM24	3–4	7·5–9·5	1·3	0·5	0·3	3·0	17
LM25	0·1	6·5–7·5	0·5	0·1	0·2–0·6	0·1	14
LM27	1·5–2·5	6–8	0·8	0·3	0·3	1·0	9

(LM27 also contains 0·2–0·6% manganese)

* LM13 is a piston alloy. Various producers have specifications based on LM13, while not necessarily adhering to that composition.

important in the mass-production industries, where fast machining and long life of cutting tools is necessary. This alloy and ones like it are diecast in vast tonnages.

Various specifications are used in different parts of the world. The aluminium–silicon alloy known as L M6 in the U.K. is referred to as A413·2 in the U.S.A., D1 V in Japan and 4261 in Denmark and Sweden. In recent years international nomenclatures have been agreed to indicate the composition of the alloys. Under the I S O specification D I S 3522, the aluminium–silicon alloy, is designated as Al Si12. The alloy with about 8 per cent silicon and 3 per cent copper, mentioned above, has an impurity allowance for iron, as will be seen from the table on page 158 and it has the I S O specification Al Si8 Cu3 Fe. This is logical but it may take a little time before foundry foremen in conversation refer to such a 'mouthful' instead of the easier L M24 or the very descriptive 380.

WROUGHT ALUMINIUM ALLOYS

As wrought aluminium products include sheet, foil, rod, bar, wire, forgings, extrusions and tubes, they cannot be classified as simply as the casting alloys. Code letters and figures indicate the form of the product and the type of heat treatment that has been given. The wrought aluminium alloy compositions are classified in eight groups, according to the main alloying element. Each has four integers and, when referring to the constituents, they are stated thus, 1 XXX, 2 XXX and so on. The first group covers pure aluminium, 1050, and a commercial grade, 1200. The remainder are shown in the table below.

Table 14

Series	Main alloy constituent
2 XXX	copper
3 XXX	manganese
4 XXX	silicon
5 XXX	magnesium
6 XXX	magnesium and silicon
7 XXX	zinc
8 XXX	lithium and miscellaneous other additions

In the 2 XXX range the alloys in the heat-treated condition have mechanical properties similar to those of mild steel: good stiffness and fatigue resistance but poor resistance to corrosion and they are often clad with pure aluminium, as will be described on page 165. Only a limited amount of manganese, up to about 1·5 per cent, can be alloyed with aluminium, so there are only a few compositions in the 3 X X X group; they have moderately good strength but excellent workability and good corrosion resistance.

The 4 X X X alloys containing silicon are suitable for wire used in welding and a few other applications. However the aluminium–silicon alloys provide the 'backbone' of the aluminium casting industry. The 5 X X X group contains magnesium as the major element, but some alloys also include small amounts of manganese. They have good strength, good weldability and corrosion resistance, and they are used especially for marine applications. The 6 X X X group containing magnesium and silicon can be heat treated; these alloys possess good weldability, formability and resistance to corrosion.

The 7 X X X group, containing up to about 6·5 per cent zinc, with about 2 per cent of copper and magnesium, and small amounts of manganese, chromium and zirconium, are among the strongest light alloys. When heat treated they have a strength of up to 600 N/mm² and are widely used in the aircraft industry. The last group of alloys containing lithium will be discussed on page 281. The compositions of some wrought alloys for aircraft are given on page 165.

Although tinned steel cans are still used extensively, the production of all-aluminium cans has grown rapidly, starting in the U.S.A. In 1980 about 20 per cent of soft drinks and beer cans in Britain were all-aluminium; by the end of 1984 the proportion had increased to about 50 per cent and has remained at that level. An alloy with one per cent of each of magnesium and manganese is employed. Food cans are made with alloys containing 2 to 4 per cent magnesium.

The easy-open tops of cans are blanked from an alloy with about 4·5 per cent magnesium. It is rolled into strips to a thickness of 0·33 mm. Each top is V-grooved to form the tab which will be pulled off the can. The savings in transport costs from the use of aluminium cans are considerable. A bottle weighs about 230 grams; tinplate cans of comparable capacity weigh between 35 and 40 grams but an all-aluminium can weighs only about 20 grams. The cans are made by a 'drawing and ironing' process in two parts – a body and a top. A disc of metal is stamped into a cup shape; the walls are then further pressed or 'ironed'. Productivity is impressive and recent developments are continually improving production.

From the point of view of scrap reclamation there is much to be said for the use of a single metal. Tinned steel bodies with aluminium tops combine three metals; they can be used as ferrous scrap, where the small percentages of tin and aluminium are not harmful, but they are not suitable as a source of aluminium. The all-aluminium can is ideal for reclamation; in the U.S.A. publicity has exploited the idea of keeping America tidy by recycling aluminium cans and in some places people are paid to return them. They are achieving recycling rates of over 50 per cent, representing more than 600 000 tonnes of used cans reclaimed each year. Sweden does even better, with a reclamation rate of about 70 per cent.

AGE-HARDENING

In the early part of the twentieth century, Dr Alfred Wilm, a German research metallurgist, was investigating the effect of additions of small quantities of copper and other metals to aluminium, in the hope of improving the strength of cartridge cases. He tried various combinations and different forms of heat treatment and, more or less by accident, an extraordinary discovery was made.

An aluminium alloy containing 3·5 per cent copper and 0·5 per cent magnesium was heated, then quenched in water and tested. The results were not particularly impressive. A few days later, some doubt was expressed about the accuracy of the tests, so pieces from the same batch were tried again. To Wilm's surprise the hardness and strength were much higher than the values already obtained; this led him to perform a series of experiments on the effect of storing the alloy for different periods after heating and quenching. The strength gradually increased to a maximum in four or five days, and the phenomenon became known as 'age-hardening'. In 1909 Wilm gave to the Dürener Metallwerke at Düren sole rights to work his patents; hence the name 'Duralumin'.*

The Zeppelin engineers of eighty years ago realized that the design of airships might be revolutionized by the use of this light, age-hardened alloy. The rolling of the alloy into sheets and strips presented a number of problems, but intense efforts were made so that most of the obstacles were overcome, and during the First World War large amounts of age-hardened aluminium alloy, strip, sheet, girders, rivets, and other parts, were used, first for Zeppelins and afterwards for other types of aircraft.

Before 1914 the National Physical Laboratory in Britain had commenced an investigation to see whether age-hardening could be induced

* Although 'Duralumin' was the name given to the age-hardening alloys developed by Wilm, the title is now a patented trade name for a range of alloys belonging to British Alcan Aluminium Ltd.

in other light alloys and particularly in any alloys which would remain strong at the temperature at which an aircraft piston has to work. This led to the discovery of 'Y-alloy' containing 4 per cent copper, 2 per cent nickel, and 1½ per cent magnesium. The strength of this alloy, now called L35, can be increased by 50 per cent by age-hardening and it can be heat treated in the cast or wrought conditions. It subsequently provided the basis of a number of special alloys, including RR58, which was first developed by Rolls-Royce in association with aluminium producers. This alloy was later specified as 2618A, containing 2·5 per cent copper, 1·5 per cent magnesium, 1·2 per cent iron and one per cent nickel. It was used for the fuselage covering of Concorde.

THE MECHANISM OF AGE-HARDENING

Metallurgists all over the world attempted to explain the mechanism of age-hardening, but it was not until some years after its discovery that the problem was even partially solved, although hardening by quenching and ageing was used extensively in the meantime. Metallurgical detective work in revealing the causes of age-hardening was slow because the instruments then available, including the metallurgical microscope, were not sufficiently sensitive to follow the minute changes in the internal structure of the alloy. Indeed, for many years it was not certain which of the impurities or added metals was responsible for age-hardening. This was partly because, even when viewed under the microscope, there was no visible difference in the structure of the treated alloy before and after ageing. Some changes in structure could be detected by X-rays but it was not until the electron microscope began to be available that it became possible to explain the mechanism of age-hardening.

When the solid alloy with 4 per cent of copper is at a temperature of 500°C, the copper enters into solid solution in the aluminium (see page 50). At room temperature, however, aluminium can hold less than half of one per cent copper in solid solution. When this alloy is quenched in water from 500°C down to room temperature, the copper does not immediately come out of solid solution.

During the lapse of some days after the alloy has been quenched, the copper atoms, which can no longer be held in solid solution, are forced to move or diffuse among the aluminium atoms to form minute areas containing higher amounts of copper than the average. These areas ultimately form the intermetallic compound $CuAl_2$ (page 52). This delayed effect may be compared with the behaviour of some kinds of home-made jams in which sugar crystallizes out, apparently of its own

Table 15

Conditions of aluminium–copper–magnesium–silicon alloy	Brinell hardness (approximate)
1. As quenched	60
2. Quenched and kept at room temperature (i.e. naturally aged)	120
3. Quenched and reheated to 175°C (i.e. precipitation treated)	150

accord, when the jam is left to stand for a long time. The reason is that when the jam is hot it will hold more sugar in solution than when cold; the sugar does not crystallize immediately the jam is cold but becomes apparent several months afterwards.

All this describes *what* takes place inside the alloy. *Why* the local aggregations of $CuAl_2$ cause age-hardening to occur was controversial for more than fifty years after Wilm's discovery. Gradually it has been realized that the hardness and strength of metals is related to resistance to slip of the crystals and increases in lattice strain. The researches on 'dislocations' described on page 62 have all added to the evidence that hardening is caused by formation of areas or zones of copper and aluminium atoms, which hinder further slip and distortion by preventing the free movement of dislocations (*Plate 24a*).

PRECIPITATION TREATMENT

The previous discussion referred to an aluminium–copper alloy in which the intermetallic compound $CuAl_2$ is precipitated at room temperature and thus increases the hardness of the alloy. The hardening can be hastened and intensified by heating the quenched alloy at about 175°C for a few hours. This is known as 'precipitation treatment'. Furthermore if magnesium and silicon are also present, a compound of these two elements, Mg_2Si, is formed which takes part in a similar process of hardening to that occurring in ageing. When the aluminium–copper–magnesium–silicon alloy is quenched and reheated to 175°C, particles of Mg_2Si are precipitated in addition to $CuAl_2$. This alloy then becomes harder than if aged at room temperature, as is shown in *Table 15*.

This type of hardening has since been discovered in alloys of other metals, such as magnesium, copper, zinc, tin, lead, and iron. Many of these will not harden by ageing at room temperature. The process involves, first, raising the alloy to a fairly high temperature at which a proportion of the appropriate alloying element goes into solid solution; this is known as 'solution heat treatment'. After a rapid quench, the alloy is precipitation treated to make the dissolved element come out of solution in minute agglomerations of such a size that maximum hardness is reached.

Alloys which age at room temperature are classed as age-hardening alloys, while those requiring precipitation at higher temperatures are called precipitation-hardening alloys. It might be noted that the essential difference between the precipitation treatment of such alloys and the heat treatment of steel is that steels attain their maximum hardness by quenching, while tempering usually *reduces* hardness. The precipitation-treated alloys, as quenched, are comparatively soft, but precipitation treatment *increases* the hardness.

The process of age-hardening aluminium alloys in aircraft factories introduced the refrigerator into metallurgical practice. When 'Dur-alumin' was first used, a part such as a rivet had to be driven almost immediately after quenching, otherwise age-hardening would commence and the alloy would become too hard to work. It was found that if, after quenching, the alloy was stored in a refrigerator at about 15°C below zero, the age-hardening change was slowed down and the rivets could therefore be stored at that low temperature until they were required.

ALUMINIUM IN AIRCRAFT

There have been aircraft in which most of the fuselage was of wood – the Mosquito, for example. Gliders, which at first were of wood and canvas, are now made of non-metallic materials such as glass fibre. Advanced military aircraft contain a large amount of titanium but most commercial aircraft rely on aluminium and its alloys for over 70 per cent of their weight. The choice of alloy depends on the stresses to be encountered, the temperature at maximum speed and the corrosion resistance require-ments. *Table 16* shows some of the alloys in modern aircraft. The principal alloying elements are shown; where small percentages of other metals are included they are shown with a single figure, indicating maxima.

Alloy 7075 is used for the Airbus frame and tail, 2024 for the surface skin, 7010 for the wing spars and wing box, 7150 for the top and 2024 for the bottom wing-surfaces.

A large proportion of the weight of an aircraft may consist of heat-treated aluminium alloy in the form of thin sheets which cover the wings

Table 16. Some aluminium alloys in aircraft

Specification	2024 %	2618 %	7010 %	7075 %	7150 %
copper	3·8–4·9	1·8–2·7	1·5–2·0	2·0–2·6	1·9–2·5
manganese	0·3–0·9	0·25	—	0·1	0·1
magnesium	1·2–1·8	1·2–1·8	2·1–2·6	1·9–2·6	2·0–2·7
zinc	0·25	0·15	5·7–6·7	5·7–6·7	5·9–6·9
zirconium	—	0·20	0·15	—	0·08–0·15
titanium	—	—	—	0·06	0·06

and fuselage. Some of the alloys in the 2 XXX range, though very strong, do not have good resistance to corrosion. To provide the necessary protection, and thus to retain the strength, a process known as cladding is applied. A thin layer of pure aluminium is rolled on to each side of the alloy sheet, making a 'three-ply' metal. This is such an efficient combination of protective outer coating with strong alloy centre that clad sheets can be exposed for over five years to the continuous action of a salt-spray test without deterioration through corrosion. The thickness of the aluminium coating varies according to the gauge but as a general rule the total thickness of the aluminium, front and back, amounts to 10 per cent of the thickness of the sheet. The rolled clad material is heated to about 500°C, quenched and thus brought to the soft condition. It is then pressed or beaten into the required shape. After that the clad metal is aged in a temperature-controlled furnace, where the alloy core of the clad sheet increases in hardness and strength due to precipitation-hardening.

Many of the Boeing aircraft operated by American Airlines make a feature of their silvery appearance. This provides scope for clad alloys to be used. Other airlines including many in Europe colour their aircraft, so the sheets of aluminium alloy are anodized and then painted; cladding is not necessary.

Weight reduction is vital, even if it entails extra cost of manufacture. Many construction features of aircraft begin as solid bars, often 150 mm thick, which are then machined to form the complex shapes of the structural members. 90 per cent or more of the metal has to be removed but the one-piece component does not require rivetting and it can be made to a specific thinness or thickness as its position requires.

ALUMINIUM IN SHIPPING

The weight of a ship's superstructure can be halved if aluminium is used instead of steel. The advantages of aluminium in improving stability and in increasing the size of superstructures were realized soon after 1945. Aluminium was used first in small vessels, including lifeboats; among the landmarks in the development of aluminium in ships are SS *United States* in 1952, then *Oriana*, *Canberra* and *France*, each of which contained about a thousand tonnes of welded aluminium alloy in their superstructures. Other industries contributed to the development of aluminium: welding of aluminium became much more practicable when the argon-arc welding process was employed.

Confidence in the use of aluminium led to the choice of aluminium–magnesium alloys in the four major decks above the steel portion of the *Queen Elizabeth 2*. The additional cost of the alloy was about £500 000, the basic cost of the ship £25 million; so the increased cost of using aluminium was about 2 per cent, justified by the increased revenue potential from the expanded carrying capacity.

Although aluminium alloy structures may cost twice as much as steel, the higher capital cost of an aluminium cabin-cruiser, bus or railway coach may be offset by reduced running charges.

ALUMINIUM IN AUTOMOBILES

The mid-1950s saw many developments for aluminium alloys, leading to the achievement of the diecast cylinder block that featured in seven American models. Economic considerations, including the cost of placing iron liners in the die before casting, caused these projects to be discontinued, though a great amount of experience in die design and manufacturing had been gained, ready for future progress.

One cylinder block, for the Chevrolet Vega, was diecast in a specially developed alloy with 17 per cent silicon and 4 per cent copper which, after a chemical etching treatment, was so abrasion-resistant that it was not necessary to insert the iron cylinder liners which are necessary when ordinary aluminium casting alloys are employed. The Vega block continued in production till 1976 and was then discontinued, but a similar 'hypereutectic' aluminium–silicon–copper alloy is now used for Porsche cars. The $4\frac{1}{2}$ litre straight-eight Porsche 928 has a cylinder block in hypereutectic alloy. As with the Vega, the Porsche block is etched to bring the hard silicon-rich crystals into relief so that it is capable of withstanding abrasion from the pistons. The blocks for the

928 and the Porsche 944 are cast by the low pressure diecasting process (page 72). As often happens a development in one area leads to progress elsewhere. The American Beaird–Poulan chain-saw, now being produced by the million, includes a diecast engine in the hypereutectic alloy. It is used extensively in several American air and Freon compressors, master brake cylinders and rotary engine housings.

In Europe the hypereutectic alloy has been employed on a much smaller scale but diecast cylinder blocks in the conventional alloys feature in many cars made in Italy and France. The cylinder blocks are designed so that the iron liners can be lifted out for reworking or replacement, so the production problems are less obstructive than those associated with the original American diecast blocks or the later blocks in the hypereutectic alloy.

Other impressive achievements took place in the 1970s, particularly in America, in the use of aluminium in automobiles, including the complex components of automatic transmission systems. Aluminium alloy wheels began as status symbols in sporty cars but with growing efficiency of the low pressure diecasting process and a well-established code of practice for quality control, aluminium alloy wheels are now produced on a large scale.

The swing from brass and copper to aluminium in automobile radiators was pioneered in Europe in the mid-1970s. During the past decade aluminium radiators and other heat-exchangers have been perfected, so that now over 80 per cent of the European market is occupied by aluminium. The most recent developments have been by Ford, where aluminium radiators are established in the Escort, Orion and Sierra ranges. *Plate 16* shows a selection of aluminium alloy components of radiators and other heat-exchangers in some modern automobiles.

The lightness and strength of aluminium alloys was the prime reason for the change. Before advantage could be taken of these properties a great deal of research had to be carried out on joining processes and alloy selection. The automotive and aluminium industries concentrated on two methods of assembly: mechanical and brazing. The former is the simplest and cheapest but is generally applicable only to engine capacities of 1600 cc or less. Larger engines require a more efficient, brazed assembly that occupies less space than a mechanically joined one. Elaborate and expensive equipment is needed for brazing processes using vacuum and flux brazing methods such as the Nocolok process, developed by Alcan. Numerous alloys have been developed for specific applications; the most common are in the 3XXX series, with manganese as the main alloying element, but both commercially pure aluminium and the alloys in the 6XXX series that contain magnesium and silicon are used to provide good combinations of strength and corrosion resistance.

During the late 1970s, in search of wider markets, the aluminium producers had been hoping, perhaps praying, for a development akin to the use of aluminium in ships or automobile cylinder blocks that had been highlights of the previous twenty years. Their prayers were answered when it was decreed that American cars must be made lighter with the object of decreasing fuel consumption by at least a third. In addition to reduction in the size of their cars, American manufacturers began a long-range fuel-saving programme that required more aluminium components to replace cast iron. For example, Ford substituted cast aluminium intake manifolds for cast iron on about 50 000 of their V8 engines, each aluminium casting weighing only 7 kg compared with the 22 kg cast iron equivalent. Other likely uses of aluminium include brake master cylinders, brake drums, bumpers and cylinder heads. Now that greater emphasis is being put on weight-saving, the aluminium cylinder block will be required once more in America, using some of the experience that has been gained in Europe. There will probably be great competition between automated sand casting, pressure diecasting, gravity diecasting and low-pressure diecasting.

Such developments are expected to increase the amount of aluminium in American cars from about 50 kg per average car at present to twice that amount during the 1990s. European cars may not show such a great increase because already they are lighter and more economical in fuel than their American big brothers, but the exercise of converting to aluminium will certainly influence European practice. The future prospects for a still wider use of aluminium in automobiles are less certain. It has been suggested that in the 1990s, much motor bodywork will be in aluminium; if that takes place, making the weight of aluminium in a car amount to about 30 per cent, the tonnage will be immense.

Even without the possible use of aluminium in bodywork the new developments described above will increase American production by 500 000 tonnes per annum. Increases in other countries will be significant, even if they do not reach the American tonnage. During the years when the aluminium producers were suffering from over-capacity and when profitability was not very satisfactory, capital expenditure and the installation of new plants slowed down. Now the challenge will be to make substantial increases in manufacturing capacity to cope with the new uses of the metal.

14

COPPER

A primitive encampment at Çatal Hüyük, near Ankara, may have been
the site where man first discovered small globules of native metallic
copper, about nine thousand years ago. These could be hammered into
useful tools or decorative beads or pins, some of which have been found
by archaeologists. The metallurgical process of smelting was first
discovered two or three thousand years afterwards, probably near Lake
Van in eastern Turkey, where there are many outcrops of copper-bearing
minerals. A fire burning above one of these would convert the mineral
into copper, to be discovered in the ashes. By about 3500 B.C. metal-
smelting was known in Egypt, Cyprus and other parts of the Middle
East; bronze had been made by smelting mixed ores of copper and tin.
Bronze weapons dating to 1800 B.C. have been found in Britain; ample
ores existed in Wales, Cornwall and Cumberland. The art of copper-
smelting remained in Britain for over three thousand years and during
the nineteenth century most of the world's requirements of copper were
produced in Swansea.

Until the Industrial Revolution, copper was regarded as a rather
splendid metal, suitable for making into bronze cannons, bells and
massive cathedral doors. Another early use of copper was for the anti-
fouling sheathing of ships' bottoms in the Napoleonic wars. Freedom
from fouling played a significant part in Nelson's victories. The
following list, beginning 26 years after Trafalgar, shows some highlights
in the use of copper in electrical engineering, linked with the career of an
Irishman – who saw the possibilities of the metal and who was partly
responsible for the mining of copper ores to meet the demand.

1831 Michael Faraday discovered the principle of electro-magnetic
     induction, which made possible the development of electric motors
     and dynamos.

1856 A young Irishman, Marcus Daly, left his home in Ballyjamesduff, to seek his fortune in America.

1861 Antonio Pacinotti invented the ring winding system, using copper, which made the electric dynamo a practical proposition. His work was published in an obscure Italian journal, but received little attention.

1862 Daly got a job as foreman of a silver mine in Nevada and from that time became well known as a mining engineer and manager with an uncanny ability to assess the potential values of metal ore deposits.

1866 A telegraph cable was successfully laid across the Atlantic.

1870 Zenobe Gramme rediscovered Pacinotti's invention and the dynamo began to be developed.

1875 Michael Healey, prospecting for silver, staked a claim on a hill in Montana. He remembered a newspaper editorial which had said 'General Grant will encircle Lee's forces and crush them like a giant anaconda'. Healey gave the memorable name to what was later to be called the richest hill on earth.

1876 Alexander Graham Bell transmitted speech by a copper telephone wire.

1878 Thomas Alva Edison produced his incandescent electric lamp.

1879 The working of an electric dynamo was demonstrated at the Berlin Exhibition, where a small electric locomotive pulled three cars containing twenty passengers.

1880 Michael Healey had now staked several silver claims near Anaconda but was needing more capital to expand them.

1881 Marcus Daly, now a prosperous mine manager, met Healey who offered him a share of the Anaconda property. Daly at once began to deepen and develop these silver mines, which often struck irritating outcrops of copper ore, which nobody particularly wanted.

1882 Edison opened the Pearl Street generating station in New York, to supply electricity for 5000 lights.

1882 Daly discovered rich copper ore at Anaconda and had a hunch that this metal was worth developing.

1882–1884 Daly shipped 37 000 tons of rich copper ore to Swansea for smelting.

1883 Daly began to operate a copper smelter in Anaconda. This was in full swing by 1884.

Thus a man of energy and vision was in the right place at the right time and within a few years the rapid growth of the use of electricity caused a tremendous surge in the production of copper. Apart from the discoveries in the U.S.A., copper ores were mined in other parts of the world and smelted in South Wales but eventually the more economical way of smelting near the mine was developed (as Daly had done) and Britain lost her proud position as centre of the copper industry. The U.S.A. is still the largest producer of copper, followed by the U.S.S.R., Chile, Canada, Zambia and Zaire, but recently mines have been opened up and smelting plants built in Australia, China, Poland and the Philippines.

Copper has a higher conductivity of heat and electricity than any other substance except silver. The pure metal is ductile and malleable and can be rolled into strips less than 0·25 mm thick, or made into foil only 0·02 mm thick; drawn into wires less than 0·02 mm in diameter; pressed, forged, beaten or spun into complicated shapes without cracking. Such ductility is also possessed by several copper alloys, notably brass.

Copper and its alloys have attractive appearance and colour, ranging from red in the pure metal to ochre, gold, yellow or white in its various alloys. They can be cast with ease and with beautiful results, as exemplified by the bronze castings that have been made during the last five thousand years. Copper and most of its alloys can be joined by such processes as soldering, brazing and welding. It is resistant to many forms of corrosion and when, with the lapse of time, copper roofs become tarnished by the atmosphere, the result is an attractive green patina known as verdigris.

About nine million tonnes of copper are produced each year, third in tonnage to iron and aluminium. Nowadays, copper has to be extracted from ores containing about one per cent, or less, of metal. Thus, high labour and energy costs are involved in converting the ores into metal, and urgent efforts are being made to discover profitable ways of obtaining copper from weak ores or from old mine dumps, such as the solvent extraction process, described on page 18.

COPPER AS CONDUCTOR OF ELECTRICITY AND HEAT

Over a hundred and fifty years ago William Cooke and Sir Charles Wheatstone put a copper telegraph wire on a section of the London and North Western Railway, between Euston and Chalk Farm. Soon afterwards attempts were made to lay telegraph wires under the sea and in 1850 the brothers Jacob and Watkins Brett laid a single copper wire, covered with gutta-percha, across the English Channel. The wire broke after being in operation for only one day, but it was replaced a year later with an armoured cable, having a four-wire stranded copper conductor which proved to be satisfactory. Then in 1857 came the first attempts to connect a cable across the Atlantic, culminating in success in 1866.

Copper has a high electric conductivity, second only to silver. About half the metal's output is used in the pure form for electric-current carrying, ranging from massive underground cables to thin wire for domestic equipment. It is also used for bus-bars, switchgear and the components of transformers, dynamos and motors.

Long-span overhead cables require high conductivity but the metal must also be strong to support its own considerable weight and to withstand additional stresses due to the effect of wind and the accumulation of ice. To achieve greater strength the copper can be alloyed with other metals but most of them cause a substantial reduction of electrical conductivity. However one element, cadmium, allows a useful compromise to be effected. When about 0·8 per cent of cadmium is alloyed with copper, the conductivity is reduced to only about nine tenths of that of pure copper but at the same time the strength of the metal is greatly increased. For many years this copper–cadmium alloy was used for overhead cable spans, but steel-reinforced aluminium, which was discussed on page 155, has now supplanted copper–cadmium in most situations. However copper–cadmium alloy is used for contact wires in railway electrification. The heat conductivity of copper is illustrated by its use for the tubes and water cylinders in domestic water heating; here its resistance to corrosion is an extra bonus. The metal is also used for flat plate collectors in solar heating apparatus.

COPPER AND OXYGEN

Even if copper were initially produced free from oxygen, the operation of casting the molten metal would normally result in a pick-up of oxygen from the atmosphere, forming cuprous oxide. If the oxygen content is above about 0·2 per cent, this causes the copper to be brittle. On the other hand in the absence of oxygen, hydrogen is absorbed by the molten metal from moisture or fuel gases; when the metal solidifies, the hydrogen, not being so soluble in the solid metal as in the liquid, is liberated and leads to a characteristic porous structure. Moreover, if both hydrogen and oxygen are present together in the molten copper there is risk of unsoundness, due to the liberation of steam, formed by interaction of the hydrogen and oxygen during the solidification of the metal. In practice, oxygen is usually kept between 0·025 and 0·05 per cent.

Oxygen-free copper is produced from cathode copper, by a process which prevents the absorption of oxygen and hydrogen from the environment. It is used for parts of radar and other electronic apparatus. Oxygen-free copper is the basis for special alloys containing elements such as zirconium, chromium and magnesium, used for components of switchgear which need a combination of good conductivity and high strength at elevated temperatures.

BRASS

The art of brass casting in Britain dates from 1693, when John Lofting, a London merchant, was granted a patent for the casting of thimbles, which previously were imported from Holland. Lofting, with three men and three boys, was soon casting 20 000 thimbles per week.

Brasses containing less than 36 per cent of zinc are ductile when cold and can be worked into complex shapes without the necessity of frequent annealing. A cartridge case, as used by NATO for a 7·62 mm bullet, is an example of the use of a brass containing 30 per cent zinc. A strip about 3·2 mm thick is cut into circular blanks about 30 mm in diameter; then each blank is pressed into the shape of a shallow cup. By a series of further pressing operations the cup is pushed and squeezed through successively smaller and smaller holes in a steel die, resulting in the walls of the cup being elongated until eventually a tube with a relatively thick base and thin walls is obtained. Further operations are carried out to the base to form the recess for the detonating cap and ensure that the cartridge fits only one type of breech. The rim of the cartridge is also softened by annealing so that it may be bent in to clip the bullet.

The brasses which contain above 36 per cent of zinc are harder and stronger than those containing less than 36 per cent. This fact has already been noted on pages 58–9, where it is shown that with up to 36 per cent zinc, the alloy consists of an alpha solid solution. Between 36 and 42 per cent, another solid solution, beta, is also present and these alloys are called alpha-beta brasses; among these, 60/40 brass is the best-known example. Although such brasses are less workable at room temperature, their plasticity is increased at high temperatures; they are usually shaped by hot-rolling, extruding, hot-stamping, casting, or die-casting.

While the alpha brasses are often straight alloys of copper and zinc, other alloying metals including aluminium, iron, tin and manganese are added to alpha-beta brasses, and their strengthening effects are in the order in which they are named, that of aluminium being greatest.

Very often lead is introduced into brass to improve its machinability.

When such a 'free-cutting' brass is machined on a lathe, the metal, which contains particles of lead, does not cling to the cutting tool in long spirals but breaks off in small chips, so a leaded brass can be machined at a much higher speed than would be possible in the absence of lead. For obtaining maximum cutting speed, about 2·5 to 4·5 per cent lead is included in the alloy, which is then used for making screw-threaded products. The presence of such an amount of lead in the brass brings about some deterioration in the mechanical properties and tends to make hot-stamping difficult, so, where brass has to be shaped by hot-stamping and then machined rapidly, only about 1 to 2·5 per cent of lead is introduced.

BRONZE

Strictly speaking, bronze is an alloy of copper with tin, but the word has come to signify rather a 'superior' material as compared with common brass. For example silicon bronze and aluminium bronze contain no tin, while manganese bronze contains only a small amount. The 'bronze' which is most familiar is the copper alloy containing $\frac{1}{2}$ per cent tin and $2\frac{1}{2}$ per cent zinc, from which old pennies, halfpennies and farthings were made, as well as the decimal one and two pence.

Before 1672 the humbler coinage was of silver but for economic reasons copper farthings and halfpennies were introduced. Later pennies and two-pences were made, in Boulton and Watt's foundry in Birmingham; these coins were so large and heavy that they were nicknamed 'cartwheels'.

The use of bronze instead of copper for coinage was one of the consequences of the French Revolution. The revolutionary atheists destroyed churches and tried to find a market for the bronze church bells. It was found that by adding an equal weight of copper to the bronze from the bells, a good coinage alloy was formed. During the years after the Revolution more and more copper was added to the bell bronze until, half a century afterwards, by trial and error, an alloy of 95 per cent copper, 4 per cent tin and 1 per cent zinc was adopted in France and then by many other countries, including Britain in 1860. The changes in the composition of bronze coinage are shown in *Table 17*.

A modification of copper–tin bronze is phosphor bronze. One type of this alloy contains 4·5 to 6 per cent of tin and less than 0·3 per cent of phosphorus, which then exists in solid solution in the bronze. Such a phosphor bronze is very suitable for springs and electrical contacting mechanisms where the alloy's resilience, non-magnetic properties, and freedom from corrosion maintain regular working.

Table 17. Composition of British bronze coins

Date	Copper per cent	Tin per cent	Zinc per cent
1860–1923	95	4	1
1923–1942	$95\frac{1}{2}$	3	$1\frac{1}{2}$
1942–1945	97	$\frac{1}{2}$	$2\frac{1}{2}$
1945–1959	$95\frac{1}{2}$	3	$1\frac{1}{2}$
1959 to date (including decimal coinage)	97	$\frac{1}{2}$	$2\frac{1}{2}$

When over 0·3 per cent phosphorus is present, the surplus separates as a hard constituent, Cu_3P; this type of phosphor bronze, which usually contains about 10 per cent of tin, is extensively used in the form of castings. It is harder than the first type and makes a good bearing-material and is also used for components which endure heavy compressive loads, such as parts of moving bridges and turntables, and rolling-mill bearings.

The alloys known as gunmetals are also copper-base alloys containing tin, and this type of alloy was known from early times, when it was used for cannons. Because gunmetal may be cast with ease, it still enjoys a wide reputation, though not for making guns. One of the classical alloys, Admiralty gunmetal, containing 88 per cent copper, 10 per cent tin, and 2 per cent zinc, has been used for marine purposes, for pump bodies, and in high-pressure steam plants. The leaded gunmetals, headed by the ubiquitous 85–5–5–5 alloy (85 per cent of copper with 5 per cent each of tin, zinc, and lead), dominate the sand-founding industry for general-purpose and pressure-tight castings.

LEADED BRONZE BEARINGS

The function of lead in a bearing bronze is to act as a sort of metallic lubricant when the oil film breaks down. In bearings for some aero-engines the leaded bronze consists of copper with about 30 per cent lead, with additions of 5 per cent or more of other elements such as zinc, tin, or nickel. The copper provides high thermal conductivity, which assists in avoiding overheating, while the presence of the minor additions improves the lead distribution and increases the mechanical strength of the bearing. These alloys are sometimes used for grinding-machine bearings; the lead acts as an absorbent, in which minute particles of grit can safely embed themselves and thereby reduce wear on the remainder of the bearing surface.

ALUMINIUM BRONZES

The aluminium bronzes, previously discussed on page 44–5, are copper–aluminium alloys to which iron, nickel, silicon or manganese can be added. They have an attractive golden colour, a strength comparable with that of steel and they resist corrosion well. Aluminium bronze is used in marine engineering for naval and mercantile shipping and for many components of small boats and pleasure cruisers. It can be used as a substitute for steel both above and below the water line and, being non-magnetic, it is used in underwater detection gear.

Products requiring extensive cold working, such as tube, sheet, strip and wire, are generally made in alloys with less than 8 per cent aluminium; they are softer and more ductile than the aluminium bronzes with between 8 and 11 per cent aluminium, which are used in the sand-cast, gravity diecast and hot worked forms. Typical examples are gear selector forks for automobiles, valve gear inserts, locking rings and dies for both compression and injection moulded plastics. During recent years aluminium bronzes have become widely used for ships' propellers, including the high strength nickel–manganese–aluminium bronze supplied for the *Queen Elizabeth 2* and an alloy containing 9·5 per cent aluminium, 4·5 per cent nickel and 5 per cent iron, for the propellers of the large crude oil carriers, discussed on pages 69–70.

OTHER ALLOYS

Although copper has high electrical conductivity it is possible, by alloying, to decrease the conductivity so much that some copper alloys have great resistance to the passage of electric current. One such copper alloy contains 13 per cent manganese and 2 per cent aluminium. These alloys also have the property that their electrical resistance does not change with variation in temperature; they are used for underfloor heating and for resistances to control the speeds of electric motors.

Other copper alloys are referred to elsewhere, copper–nickel alloys and 'nickel silvers' on page 195, and beryllium copper on page 205. Partly because copper forms alloys with so many other metals but also because it has been known for thousands of years, very many copper alloys have been in use, ranging from the modern temper-hardening copper–chromium alloys to the tin bronzes, which are as old as the art of metallurgy.

15

THREE COMMON METALS – TIN, ZINC AND LEAD

TIN

Tin is a soft, weak metal, slightly less dense than iron and with a low melting point (232°C). It is comparatively expensive. As a result of a buffer stock operation, managed by producers and consumers through the International Tin Agreement, the price of tin rose relentlessly from about £3000 per tonne in 1975 to about £10 000 per tonne in 1985. This brought low-cost producers into the market and the price fell to about £4000 per tonne. After three years at that level, the price began to rise and by the Spring of 1989 stood at £4600 per tonne. The high price of tin means that it is used mainly in situations where maximum advantage can be taken of the metal's corrosion-resistance and non-toxicity, from cars to cans, from heavy engineering to micro-circuits. Yet the man in the street probably thinks tin is cheap; one often hears, 'it's only a bit of tin'. This reflects the popular confusion of tin with tinplate. Tinplate is thin steel sheet, coated with a much thinner layer of tin. It is an excellent combination of metals because it has the strength and rigidity of steel, plus the attractive appearance, corrosion-resistance and good weldability of the tin coating. If steel alone were used for food storage the metal would rust, while tin alone would be too soft and much too expensive.

The story of tinplate's progress from an interesting invention to a mass-production industry began with Napoleon Bonaparte.

Napoleon was concerned with the problem of feeding his troops who, as every schoolboy knows, marched on their stomachs and whose efficient movement over great distances in good fighting order provided one of the shock tactics which enabled Napoleon to win so many battles. In 1795 he offered a prize of 12 000 francs for a method of keeping food fresh for long periods. The winner was Nicholas Appert of Paris, who discovered that when foodstuffs were boiled in glass bottles and immediately sealed, they would keep for several months. This led to other processes being tried and among them was a method patented by an Englishman, Peter Durand, of enclosing food in containers of iron, coated with a thin film of tin. This was the first time that tinning had been used for the preservation of food, although the art of coating iron with tin had been practised on a large scale in Bohemia in the thirteenth century when tinned iron was used for decorative articles and for parts of armour.

In the nineteenth century the manufacture of tinplate was practically a British monopoly, until 1891, when the U.S.A. imposed tariff protection to this manufacture. By 1912 the U.K. and the U.S.A. were making three quarters of a million tons each, but during the 1914–18 war, British production was halved and that of the U.S.A. doubled. In the meantime great strides had been made in the rapid production of cans, starting in 1846 when a device was invented for increasing the rate of output from six to sixty per hour. The first automatic can-making machinery was introduced during the 1880s, leading to the modern plants which produce a thousand cans per minute.

In 1939 total world production had risen to about three and a half million tonnes; there was a decline during the war years, because many tin-mining areas were in countries affected by the conflict with Japan. After the war Britain and the U.S.A. were still major producers, but other countries began to develop tinplate manufacture. Production peaked at 14 million tonnes in 1980 but fell to 11 million tonnes in 1986. In that year over 850 000 tonnes of tinplate were produced in Britain, using about 3300 tonnes of tin. In the composition of tinplate, tin occupies only 0·39 per cent.

Tinplate is made with mild steel containing about a tenth of one per cent of carbon. The steel is continuously cast, or hot-rolled into strips about a metre wide, 3 mm thick and up to 1000 metres long. Coils of this steel, weighing up to 12 tonnes, are brought from the hot-rolling mill, pickled in acid to remove scale, and prepared for cold rolling, to reduce the thickness from 3 mm down to the finished gauge, which nowadays can be as little as 0·15 mm. The cold rolling is done by a 5-stand high-speed tandem mill (page 82). The maximum speed at exit is about 1500 m per minute. After removing all traces of rolling lubricant, the steel is annealed by heating in a non-oxidizing atmosphere. Then it is given a final, very light rolling treatment without lubrication, which confers the required mechanical properties and surface finish.

Over 95 per cent of world production of tinplate is manufactured by electrodeposition in continuous automatic plants with annual capacities of up to 200 000 tonnes of finished tinplate, able to handle the strip at speeds of over 300 m per minute. Electrodeposition has almost entirely superseded the dated hot-dip process of immersing sheets in molten tin because of advantages in production speed, control of thickness and uniformity of the coating, and ability to produce a thinner deposit. It has a special advantage, that different thicknesses of tin may be applied to the two faces of the strip. The coating thicknesses in common use are in the range 0·0004 to 0·002 mm.

In about 60 per cent of tinplate production, the tin is electrodeposited from a solution containing stannous sulphate, phenolsulphonic acid and organic addition agents which ensure a smooth coherent coating. The steel strip is fed continuously through what is aptly called a 'vertical serpentine' electrolytic tinplate line. It is about 100 m long and at any one time a length of over 900 m of strip is travelling up and down in vertical loops as it is uncoiled, cleaned, pickled, plated, momentarily heated to fuse the tin, passed through a 'passivation solution', oiled, automatically inspected and re-coiled. At full speed the tinplate takes only about three minutes in passage from one end of the line to the other.

The brief fusion of the tin coating, known as 'flow brightening', gives brilliance to the coating. The passivation solution, of sodium dichromate, confers improved resistance to oxidation and staining; the oiling process is applied to ease the handling of the tinplate in can-making.

Most of the remainder of the electrolytic tinplate production uses what is called the 'halogen process' in which the tin is deposited from a weak acid solution containing stannous chloride and the fluoride of an alkali metal with an organic addition agent, and the steel passes through the cleaning and plating sections horizontally. One face is plated before the strip is looped back for plating the other face. Otherwise the sequence of operations is as described for the vertical serpentine process.

Tin's low melting point and ready ability to alloy with other metals are reasons for its good solderability, so tin and tin–lead alloys are frequently employed to coat electrical or mechanical components, which are later required to be soldered. Nearly one third of the world's output of tin is used for making solders. A vast tonnage is required for electronic and electrical uses. The rapid developments in these industries have made it necessary to devise fast, automatic methods of soldering. Printed circuits are now produced by wave soldering, in which the circuit assembly is passed across the crest of a wave of continuously circulating molten solder. In the past the seams of tin cans were automatically soldered but today they are produced either by welding the side-seam instead of soldering, or by drawing processes which produce a container with a seamless body and one integral end.

Pewter is a tin alloy: there are two compositions in general use, one with 6 per cent antimony and 2 per cent copper; the other contains 4 per cent antimony and 2 per cent copper. That alloy is gravity diecast in Germany to make some very splendid beer tankards. Also pewter candlesticks, tea and coffee serving trays, salt and pepper cruets are among the many items available. Although pewter may be regarded as an old-fashioned metal, it is benefiting from an increase in popularity resulting from the introduction of new methods of manufacturing pewterware and under the stimulus of modern design. Pewter is an easy metal to fashion and many home handymen and women have had their first experience in metal-beating with pewter for ornaments and vessels.

Bearing metals contain tin; for example, one bearing metal used in diesel engines and generators contains 90 per cent tin, 7 per cent antimony, and 3 per cent copper. At the other end of the scale some lead-alloy bearing metals contain 5 to 10 per cent tin. An important bearing alloy for motor-car engines is an 80 per cent aluminium 20 per cent tin alloy developed in Britain and now used throughout the world.

Bronze and gunmetal are other well-known alloys containing tin. The sound of church bells is due to the sonority of bronzes containing tin. Another example of the musical use of bronze is in the manufacture of cymbals, consisting of copper alloyed with 15 to 20 per cent tin; this alloy is worked into the characteristic disc shape. The manufacture of the best cymbals was for years a monopoly of Turkish craftsmen, but is now being achieved in Britain. Many organ pipes are of tin alloy, the sound being considered better than from pipes made of zinc or aluminium.

A fairly recent development has proved one of the most exciting uses of tin. The Pilkington float glass process involves running molten glass on a bath of pure liquid tin; the contents of the bath amount to anything up to 100 tonnes. Tin was chosen for this purpose because of its combination of properties: the requirements were for a supporting liquid which would have a density greater than glass, be molten below 600°C, but have a high boiling point and have virtually no chemical interaction with glass.

ZINC

The camp-fire of our forefathers led to the discovery of many metals. When oxide ores of copper, tin, lead or iron were accidentally heated in a fire, the carbon in the burning wood united with the oxygen in the metal compound, leaving a more or less pure metal to be found when the fire had died down. Zinc, however, could not have been discovered easily in this way; at bright red heat the liquid metal boils and so, although metallic zinc may have been formed in the camp-fire, it would become vapour which quickly oxidized in contact with the air, forming a cloud of white fumes, easily mistaken for smoke.

Long before zinc was known as a metal, the Romans mixed calamine, which contains zinc carbonate, with copper ores; the smelting of the two materials produced brass. Zinc was not isolated for many hundreds of years after the discovery of brass; it was first made in Sumatra and China, whence it was exported to Europe in the early seventeenth century. England was the first European country to develop the manufacture of zinc; William Champion of Bristol was smelting the metal on a commercial scale in 1738.

The metal is contained in zinc sulphide ore deposits widely distributed throughout the world. Important mines are in Canada, the U.S.S.R., Australia, Peru and the U.S.A. Deposits in Eire provide most of European needs. Zinc is often smelted or electrolytically refined in the countries where the mines are situated but large quantities of ore, usually concentrated by the flotation process (page 16), are also sent for treatment to Britain, Germany, Belgium, Japan and the U.S.A. The total world production in 1987 was about seven million tonnes, of which the U.K. consumed about 190 000 tonnes.

The original process for the production of zinc was a thermal one, and such methods still account for a fifth of the metal trade. The zinc oxide formed by roasting sulphide ore was heated to a temperature of about 1100°C with anthracite or similar carbonaceous material in banks of small horizontal fireclay retorts. Zinc was formed as a vapour, which was caught as liquid metal in condensers adjoining the furnace.

In the late 1920s an improvement to the thermal method was developed. In this – the vertical retort process – a briquetted mixture of roasted concentrates and bituminous coal is heated in a large vertical retort made of silicon carbide bricks, in which reduction can proceed continuously. This semi-mechanized plant greatly reduces the labour needed, in comparison with hand-operated retorts, but its initial cost and maintenance are greater. The daily output of a vertical retort is 8 to 9 tonnes of metal. By a subsequent process of fractional distillation, a metal with a purity of more than 99·99 per cent can be produced for the manufacture of zinc diecasting alloys and for other purposes.

The electrolytic zinc process, developed during the First World War, now accounts for about four fifths of world zinc production. Roasted concentrates are leached in sulphuric acid; and, after purification of the solution, zinc is deposited electrolytically on aluminium sheets from which it is stripped off, melted, and cast into slabs; the acid is simultaneously regenerated and used repeatedly. The purity of the metal so formed is greater than 99·95 per cent and can be maintained above 99·99 per cent when desired.

The Imperial Smelting Process

After works experiments which started in the 1940s, Imperial Smelting Corporation, at Avonmouth, announced in 1957 the successful development of a blast furnace for making zinc. This was a notable achievement which involved the simultaneous smelting of zinc and lead, compounds of the two metals normally occurring together in nature.

In the Imperial Smelting Process, zinc and lead sulphides are roasted to produce oxides which are charged in the blast furnace with proportioned amounts of coke. Preheated blast air enters the furnace through water-cooled tuyères. Slag and molten lead containing precious metals and copper from the charge are tapped from the bottom of the furnace and separated. Zinc leaves the shaft as a vapour in the furnace gas containing carbon monoxide and carbon dioxide and is shock-cooled in a lead splash condenser, the zinc vapour being absorbed by the lead. Next molten lead containing zinc in solution is pumped out of the condenser and cooled so that zinc comes out of solution and floats on the lead. The zinc layer is poured off and cast and the cooled lead is recycled to the condenser.

The invention of the lead splash condenser provided the key to the process, as its extremely rapid cooling prevented the oxidation of the zinc vapour by carbon dioxide which had frustrated earlier researchers. There are now thirteen Imperial Smelting furnaces, producing annually nearly a million tonnes of zinc and 260 000 tonnes of lead. Some of the metal is then redistilled to make high purity zinc.

Uses of Zinc

The protection of steel from rust and the making of brass and zinc alloy diecastings take over 80 per cent of the total consumption of zinc. Its other uses include rolled zinc for building and zinc powder for protective paints and alkaline batteries. Zinc oxide is an ingredient in synthetic rubber, ceramics and in the production of some zinc compounds.

Zinc sheet was adopted as a roof covering in Europe early in the nineteenth century, particularly in France, Belgium, Germany and the Netherlands, where it remains popular for long-lasting roofs, wall-cladding and rainwater goods. Many British railway stations and seaside piers have been roofed with zinc and a life of 40 years can be expected if the work is done properly.

Zinc Coating Processes

The biggest single use of zinc is for coating on steel to protect it against rust and the most important process is hot dip galvanizing in which steel articles are immersed in molten zinc. The many uses of galvanizing range from household buckets to electric power transmission guards, highway guard rails and the underbody protection of automobiles. Over the last fifty years the large-scale production of continuously galvanized steel strip has replaced individually dipped sheets. This material has a ductile coating and will stand the same amount of deformation as the steel itself.

Electro-galvanizing consists of electroplating zinc on the previously cleaned article. Although electro-galvanized coatings have a good appearance, they are applied primarily to prevent corrosion. Zinc plating by continuous process is used to protect steel strip, as well as small nuts and bolts. The coating is thinner than that applied by hot-dipping and is intended as a base for painting, to give added protection.

Zinc is coated on to metals for protective purposes in several other ways. One method consists in spraying zinc from a 'metallization pistol'; zinc wire is fed into the pistol, where it is melted by an oxy-gas flame or electric arc and, in atomized form, blown on to the article by compressed air; the action is similar to that of a scent spray. This process is used for covering ships' hulls and anchors, bridge steelwork and large tanks. Metal-sprayed coatings, subsequently painted, give very good protection to steel structures. Aluminium is also metal sprayed.

Sherardizing is another coating process; steel articles are placed in a rotating drum, together with zinc powder and fine sand. The container is then sealed and heated for several hours at about 375°C when the zinc diffuses into the steel, giving a fine-grained protective coat of zinc–iron alloy. Small parts such as springs, washers, nuts and bolts are Sherardized automatically. Small steel parts are also coated by mechanical plating. The parts to be protected are cleaned and then tumbled in a drum containing zinc dust, tiny glass balls and an activating solution. The balls hammer the zinc on to the steel surfaces, to which it becomes cold-welded. No heating is needed, so the properties of the steel are unchanged.

Zinc Alloy Diecastings

Automobile mass-production, from its earliest days, provided a market for zinc alloy diecastings. Initially simple items such as badges and door handles were made but now complex and sometimes massive parts are produced, generally automatically. Carburettors, fuel pumps, windscreen wipers, door and steering column locks are typical examples. When necessary the diecastings are electroplated with copper, nickel and then chromium.

Today zinc diecastings are used in a vast range of applications in many industries. Builders' hardware, small power tools, electrical and electronic components rely on the accuracy, reproducibility and reliability of zinc alloy diecastings. Sizes cast range from several kilograms down to a few milligrams; so far as we can ascertain, the smallest zinc diecasting is a wedge pin for watches, weighing only 1/300 of a gram.

Unalloyed zinc dissolves iron and steel, so the pure metal would not be suitable for making diecastings. When 4 per cent of aluminium is alloyed with the zinc, this effect is inhibited; furthermore, the alloy casts with ease and it is strong. However, these zinc–aluminium alloys are very sensitive to the effects of traces of certain impurities. This is in contrast to most other foundry alloys, where small amounts of impurities are far from being catastrophic. Traces of tin, lead and cadmium would cause the diecastings to become brittle. The ready availability, since the 1930s, of 'special high grade' or 'four nines' zinc (over 99·99 per cent pure) was the key to the success of zinc diecasting alloys. Zinc of this grade is obtained either by re-distilling the metal produced by smelting or directly from electrolytic refining and it enables stable, strong alloys to be produced; a small amount of magnesium (around 0·05 per cent) is added to counteract the effects of any remaining traces of impurities.

The purity of these alloys is so essential that in the 1950s diecasters in Britain, America, France, Germany, Australia and, later, many other countries evolved certification schemes under which they guaranteed to use nothing but zinc alloy of maximum purity. As a further precaution, each day's production is analysed, and the castings held in bond till each batch has been certified as being of the required purity of composition. The zinc diecasting alloys are generally known in Europe as 'Mazak', a name derived from the initial letters of the words Magnesium, Aluminium, Zinc and 'Kopper'.* In the U.S.A. the alloys are called 'Zamak'.

---

* In the early days of the industry about 3 per cent of copper was included; now the diecasting zinc alloys are either copper-free or contain about 0·75 per cent copper.

In the 1970s a further range of zinc–aluminium alloys was developed, containing 8, 12 and 27 per cent of aluminium and with the same careful control of impurity levels as in the 4 per cent alloy. These alloys are extending the range of applications for which zinc diecastings are suitable, particularly for improving strengths at higher temperatures, improved creep resistance and good bearing properties.

LEAD

The Romans' elaborate system of water distribution made it necessary to obtain large amounts of lead; Greece, Spain and Britain were principal sources of supply. For lining Roman baths the metal was cast into thick sheets. Water pipes, in lengths of about 3 metres, were made by bending a lead sheet over a rod so that a tube shape was produced. The joint was sealed with solder or with lead by the Roman *plumbarii*. Some of the lead pipes and conduits, made two thousand years ago, can be seen in museums in Rome and in such buildings as the house of Livia, wife of Augustus Caesar, on the Palatine Hill. Others have been discovered at Pompeii and Bath.

A property which makes lead easy to work is its low melting point (327°C). Its ease of casting is one of the reasons why for centuries lead was associated with the manufacture of printers' type metal; all editions of this book until 1980 were printed using an alloy containing 85 per cent lead, 11 per cent antimony and 4 per cent tin. (The present edition has been produced by computerized photo-typesetting.)

The combination of ease of melting and low strength made the ancients regard lead as the least noble of metals, and it will be remembered that in *The Merchant of Venice* the casket chosen by Bassanio was of 'base' lead. Yet notwithstanding its low rank in the family of metals, lead was used for coinage in early times because it was easy to cast and because it resisted corrosion.

The world's annual production of lead is over five million tonnes, though only about two thirds of this is newly smelted metal. Lead is so easy to melt and refine that a large amount of metal can be reclaimed from scrap batteries, sheet, pipes and cable sheathing. Oxides and impurities are removed, the alloy is brought to the required composition, pumped from the furnace and automatically cast into ingots. Apart from the scrap treatment described above, the main sources are from lead ores in the U.S.S.R., the U.S.A., Australia, Canada, Mexico and Peru.

About 100 000 tonnes of lead, as metal and oxides, is used for making car batteries in Britain, while in the U.S.A. they require over 900 000 tonnes, roughly 75 per cent of their total consumption of lead. The 20 million cars in Britain use lead–acid batteries; over 5 million per annum are made, of which about 4 million are replacements. In the past decade the electric power per unit weight in the lead–acid battery has nearly doubled. Such improvements will increase the opportunities for electric urban delivery vehicles, paving the way for electric or hybrid (battery plus internal combustion) passenger cars.

Flat-plate batteries for automobiles and lighting are made of lead alloys, containing antimony, or, more recently, calcium; lead oxide pastes are pressed into the form of grids. The manufacture of battery grids has become a highly mechanized process. They are made either by high speed gravity diecasting or by expanding continuously cast lead strip. In the U.S.A. a new technique is evolving in which battery grids are cast continuously at speeds of over 400 per minute.

Flat-plate batteries are not heavily loaded; they would have a short life if subjected to the duties of material handling, mining, railways and submarines. Heavy-duty batteries use lead–antimony alloy 'spines', located in gauntlets filled with lead oxide, giving long life under arduous conditions.

Lead has been used in building work since early Egyptian times. Nearer home there are many lead roofs which, like those in Westminster Abbey and St Paul's, have lasted for several centuries. Lead sheet is the most widely used material for flashings and weatherings under windows, chimneys and on parapet walls. In recent years lead sheet has been used increasingly as a cladding material on good quality buildings; currently about 100 000 tonnes of lead sheet are required annually for buildings in the United Kingdom. Lead sheet can be bonded to plywood and chipboard, and these laminates make effective partitions with good sound insulation in public buildings.

Lead will not dissolve completely in some other liquid metals; for example molten lead and aluminium will not mix. Alloys of lead and copper, containing between 40 and 95 per cent lead, are not fully miscible in the liquid state except at temperatures in excess of 1000°C. However, lead does form useful alloys with metals of similar low strengths and melting point. About 7000 tonnes per annum of lead–tin alloy solders are made in Britain. Apart from battery plates, mentioned above, the lead–antimony alloys are used in small arms ammunition and for ornamental castings.

Lead is rolled into sheets and it can easily be extruded into rods, pipes, and collapsible tubes. The low strength of lead makes it unsuitable for engineering where considerable stresses are encountered, and it is not surprising that while work was progressing with the strong alloy steels, aluminium, magnesium, nickel, and copper alloys, little attention was paid to the possibility of improving the strength of lead, which has only a fraction of the strength of steel. Nevertheless, there are good reasons why attempts should be made to improve the mechanical properties of any lead components which are subjected to stress.

In the past, lead has been thought of as somewhat unglamorous, though capable of doing a great deal of metallurgical donkey-work. However, to mix our metaphors, lead, like other dark horses, often brings surprises. Many modern developments of lead and its alloys have been enterprising and sophisticated.

In nuclear research establishments and in hospitals where X-rays, radium and radioactive isotopes are involved in examination and treatment, lead is used in the form of interlocking bricks made of the metal alloyed with 4 per cent of antimony. Radiography and hospital treatment rooms are constructed with sheet lead attached to other materials, such as plywood. Altogether about 5000 tonnes of lead are used in Britain each year for radiation shielding.

Lead is also used as a protective sheath or electric power cables. Although substitution by plastics has occurred, lead is still the best material for underwater applications. Lead's corrosion resistance and density makes it ideal for many applications, from tiny castings for precision instruments to multi-tonne yacht keels. Finally, about a third of the world's consumption of the metal is for chemical compounds, not only as lead oxide for batteries but in glass, ceramics and stabilizers for plastics.

16

NICKEL

The name of this metal derives from the experiences of fifteenth-century miners who suffered from arsenical poisoning whilst attempting to recover copper from ores in the hills of Saxony in Germany. Their misfortunes, and their inability to extract a satisfactory copper, were assumed to be caused by evil spirits, so they named the ore 'Kupfer-nickel' or 'Devil's Copper'. Much later the nature of the ore was explained by Axel Cronstedt in 1751; it was copper-based but contained a compound of arsenic with a new metallic element which he named nickel. The new metal appeared to be brittle and of no practical use until, fifty years later, it was shown that the brittleness was caused by carbon and sulphur which could be eliminated by additions of manganese and magnesium. Nickel was then found to be a ductile, strong and workable metal with good corrosion resistance.

In 1867 large deposits of nickel ore were found in New Caledonia, an island about a thousand miles east of Australia. In 1883, during the construction of the Canadian Pacific railway, a massive outcrop of nickel–copper ore was found near Sudbury, which remains the largest single source of nickel in the western world. Much of the world's nickel ore also comes from New Caledonia and there are supplies from Russia, Australia, Cuba, the Philippines, Botswana, Zambia, Indonesia, Greece and several sources in the American continent.

The two most widely mined nickel ores are nickel sulphide in the hard rock formations of Canada, Russia, Australia and Africa, and what are known as lateritic ores, of the iron–nickel oxide type, found near to the surface in countries within the equatorial rain belt. The deep underground mining of sulphide ores is capital and labour intensive but, because the minerals can be concentrated, the extraction and refining processes are comparatively low-cost. Furthermore some sulphide ores contain other valuable elements such as copper, cobalt, gold, silver and platinum metals, most of which are absent from lateritic ores.

Numerous lateritic orebodies were opened up during the 60s and 70s, encouraged by good nickel prices and supply shortages from the traditional producers. Mining these ores is low in labour cost and capital expenditure but, because the minerals cannot be concentrated economically by froth flotation (page 16), the whole ore must be transported in bulk, dried (it can contain over 50 per cent moisture), and then smelted in blast furnaces before a suitable concentrate for further smelting, leaching and refining can be achieved. These stages in the processing of lateritic ores are energy-intensive and the steep rises in oil prices since 1974 made many of the orebodies unprofitable. About 55 per cent of the world's nickel comes from sulphide ores and the remainder from lateritic ores.

The smelting and refining processes vary according to the type of ore. The sulphide ores containing only about one per cent of nickel are concentrated by flotation or by magnetic separation. These are followed by a series of pyrometallurgical processes that yield an impure form of nickel, as pellets or powder, which will then, if necessary, be treated by refining processes. Some ores are given a complex sequence of solvent extraction treatments, involving the skilled use of the rate of reaction and selective solubility of nickel, copper, iron and cobalt in ammonia under controlled conditions of temperature, pressure and agitation. The end product of this route is a pure nickel powder, which may then be compacted into manageable shapes.

After concentration the lateritic ores are usually smelted to produce ferro-nickel for direct use in alloy steelmaking, or the processing is continued to refine the smelter concentrate and then convert it into pure nickel rondelles by a reduction process, or into nickel squares by electrolytic refining.

One of the most efficient methods of refining nickel is the carbonyl system, developed by Ludwig Mond and Carl Langer in the latter part of the nineteenth century. They discovered that freshly reduced nickel reacts with carbon monoxide at about 50°C to produce gaseous nickel carbonyl, which can then be decomposed into nickel and carbon monoxide by heat at about 200°C. This reversible process is the basis of automated plants. Nickel oxide, produced by roasting nickel sulphide, is reduced to an impure form of nickel by treatment with hydrogen. The metal is fed into a volatilization kiln in which it meets a counterflow of carbon monoxide at 60°C. Decomposition of the nickel carbonyl gas is effected on the surface of pre-heated nickel pellets which flow continuously through a reaction chamber until they grow to a marketable size.

The present world production of nickel amounts to about 790 000 tonnes per annum. 55 per cent of this is required for stainless steels, 20 per cent for non-ferrous alloys and about 5 per cent for other alloy electroplating. The remainder is used in nickel cast irons, nickel alloy castings, catalysts and other applications.

USES OF PURE NICKEL

Nickel is a metal similar to iron in some of its properties. It has a melting point of 1455°C, nearly as high as that of iron; it has slightly lower strength and hardness and is magnetic, though to a lesser degree than iron. In contrast to iron, however, nickel is strongly resistant to corrosion; for example, it is not corroded by alkalis or chlorine, and is used for the construction of plant making these materials.

The best-known use of pure nickel is in electroplating, where it forms the corrosion-resistant layer underneath chromium plate. This durable system of decorating and protecting base metals is widely used in domestic hardware and all manner of fixings, trim, knobs and hooks. After preliminary cleaning processes, the articles to be plated are suspended in vats containing a solution of nickel sulphate with small additions, for example, nickel chloride and boric acid. Anodes of pure nickel, or titanium-mesh baskets containing nickel pieces are also suspended; these are the anodes and as nickel is deposited they replenish the concentration of nickel in the electrolyte. Low-voltage direct current flows from the anodes via the electrolyte to the articles being plated, which act as cathodes, and nickel is deposited on them. The contents of the vats are held at about 40°C; nowadays the whole sequence of operations is automated and computer-controlled.

Modern plating solutions deposit nickel at high rates and provide a bright, smooth finish that needs little polishing. The micro-cracked chromium finish now regularly applied to nickel plated articles has further enhanced the durability of this versatile surface finish. By making the surface of certain plastics electroconductive, nickel and chromium plating can also provide a hard-wearing finish to a variety of moulded products.

Electroforming is another use for plated nickel in which a thick layer of the metal is deposited on a shaped wax mandrel. When sufficient nickel has been applied, the wax is melted out, leaving a cavity or mould which is a perfect replica in reverse of the mandrel. Many moulds for plastic components and all the gramophone disc-stampers which produce records are created by this method.

There is much interest in the use of hydrogen as a source of energy.

The gas can be compressed and supplied as liquid but a more efficient way of storing hydrogen is by means of an intermetallic compound of nickel with the rare earth metal lanthanum. This absorbs hydrogen so effectively that a given volume of the compound will hold about twice as much hydrogen as the same volume of liquid hydrogen. The hydrogen can be extracted from the compound by reducing the pressure or increasing the temperature.

STAINLESS STEELS

These have been discussed, at some length, on pages 148–50. The martensitic and ferritic stainless steels are based on additions of chromium but the largest group, austenitic stainless steels, contain nickel and they represent the predominant single use of the metal. Nickel stabilizes the austenite and gives the steel added ductility, toughness and corrosion resistance. Recently a process has been developed to provide a rich range of colours on austenitic stainless steels for decorative purposes on buildings and shop fronts.

NICKEL STEELS

In 1889, James Riley presented a now historic paper to the Iron and Steel Institute of Great Britain, giving a survey of the possibilities of nickel steels. Four years previously nickel steel armour plate had been produced in France, then in Italy and Britain. Tests showed that a steel containing 3 per cent of nickel developed remarkable projectile-resisting qualities, coupled with tremendous toughness, making it suitable for armour plate. Riley's report, which shook the armament and engineering world, covered a wide range of steels with from 1 to 49 per cent nickel, and he announced that, by hardening and tempering, the almost unbelievable strength of 96 tons per sq. in. (1500 N/mm²) had been achieved. He confirmed that steels rich in nickel are practically non-corrodible and he concluded, 'I find some difficulty in not becoming enthusiastic, for in the wide range of properties possessed by these alloys, it really seems as if any conceivable demand could be met and satisfied'.

From that time onwards nickel alloy steels became vital materials and until reliable ways of producing similar properties in other alloy steels were found the nickel steel was ubiquitous. These steels are still required for strong components of large cross-section and where exceptionally high strength is needed.

CRYOGENIC STEELS

The modern interest in handling liquid gases at sub-normal temperatures has renewed the need for nickel in alloy steels because it lowers the impact-transition temperature at which steels become brittle. A relatively small addition of nickel to carbon steel can make an appreciable improvement but, for really low temperatures, steels containing 9 per cent nickel which can be welded on site are needed.

MARAGING HIGH-STRENGTH STEELS

In the late 1950s a number of new super-alloy steels were developed containing 18 per cent nickel, 8–12 per cent cobalt, 3–5 per cent molybdenum and small amounts of titanium and aluminium. These highly-alloyed steels can be forged, shaped and machined in the un-hardened condition and then brought up to very high tensile strength by a relatively low temperature heat treatment which does not distort or crack a component of complex shape. Components in other steels of similar strengths have to be fabricated with difficulty in the fully hardened condition. Maraging steels have been used in aircraft undercarriages, portable military bridges and for complex tooling such as that required in diecasting, where any distortion caused by heat treatment could not be corrected easily. They are called maraging steels because their structure is similar to that of martensite, discussed on page 137, and because they can be age-hardened.

NICKEL CAST IRONS

In 1925 it was discovered that the addition of up to 3 per cent nickel improved the toughness and strength of grey cast irons and did much to avoid unmachinable areas of high hardness. These nickel cast irons became popular wherever manufacturers needed consistent properties in components such as engine castings. Nowadays these low-nickel cast irons are not often specified; to quote one expert, 'small amounts of nickel in cast iron are neither harmful nor beneficial'. On the other hand the austenitic cast irons, containing at least 13 per cent nickel, chromium, silicon and manganese, comprise a series of corrosion-resistant irons frequently referred to as Ni-Resist irons. They are widely used for body and impeller castings for large pumps handling sea water for cooling power station condensers or for desalination plant feed water. The most highly alloyed of these irons is non-magnetic and is used for stator plates in large alternators and for mine-sweeper winches. The martensitic cast irons, containing 4 per cent nickel, with $2\frac{1}{2}$ per cent chromium and silicon, are intensely hard and are used for ball mill liners and for pump bodies handling abrasive slurries.

HEAT-RESISTANT ALLOYS

A typical high-nickel, heat-resistant alloy is based on 80 per cent nickel, 20 per cent chromium, plus small additions of rare-earth metals to inhibit crystal growth at high temperature. Such an alloy is used for heating elements, ranging from the domestic electric toaster to giant electric furnaces. Like all heat-resistant alloys it resists destructive oxidation at high temperatures, e.g. 1000°C; it is ductile enough when cold to enable it to be manufactured into wire, rod, sheet and tube strip which can then be fabricated by normal means and joined by welding.

Developments of this 80–20 alloy led to the range known as Nimonics. They are given serial numbers to indicate their suitability for operation at progressively higher temperatures. In Nimonic 75 the 80–20 binary alloy is strengthened by titanium carbide. Nimonic 90 is a nickel–chromium–cobalt alloy, to which small additions of titanium and aluminium are made. Other alloys in this range include Nimonics 105 and 115, containing molybdenum.

There are also several nickel–chromium–iron alloys, including Inconel, with about 77 per cent nickel, 16 per cent chromium and 7 per cent iron, used for furnace components and heat treatment equipment. Another heat-resisting alloy is Incoloy, with 32 per cent nickel, 20 per cent chromium and the balance iron. Both Inconel and Incoloy can be modified by the addition of molybdenum, aluminium, niobium and some of the rare earth metals. They have excellent resistance to corrosion and good strength at normal and elevated temperatures.

SUPER-ALLOYS

Since the middle of World War Two the ever-increasing demands made on aero-engines have been the motivation behind the search for new materials for service at high temperatures. These demands have been met by the development of super-alloys, many of them based on nickel. Their largest use is in the gas turbine engine, which ingests air from the atmosphere, compresses it several times, adds fuel and burns the mixture, producing turbine inlet gases in the temperature range 700°–1400°C. Super-alloys are used in components such as blades, vanes, discs and ducts.

There is a great variety of nickel-based super-alloys; most of them contain 10 to 12 per cent chromium, up to about 8 per cent aluminium and small amounts of boron, zirconium and carbon. Some of the alloys contain traces of one or more of cobalt, molybdenum, niobium, tungsten and tantalum. In addition to aircraft, industrial and marine gas turbines, these alloys are used in nuclear reactors, power generators, rocket engines and space vehicles.

CORROSION-RESISTANT ALLOYS

Nickel–chromium–iron and iron–nickel–chromium alloys feature in heat-resistant alloy systems where resistance to corrosion at high temperatures is the main objective, combined with good high-temperature strength. A familiar application is the metal tube which encloses the heating element of electric cooker hobs. This alloy has to resist attack by numerous kinds of spilt cooking fluids as well as having to withstand mechanical damage by the cooking utensils. A more extreme application is the steam-generator heat exchanger in pressurized water nuclear reactors. Other applications occur in chemical plant furnaces, flame tubes and burners. Even more complex nickel–chromium–molybdenum alloys are used to contain highly corrosive chemicals.

LOW-EXPANSION ALLOYS

These include the nickel–iron series of alloys which have a controlled rate of thermal expansion or even a zero coefficient of expansion; well-known trade names are Invar and Nilo. One application is for clock pendulums but the greatest use is for bimetallic strips, comprising two alloys of different coefficients of expansion, which deflect when heated and operate temperature-controlling valves and switches. Other nickel alloys are used in cathode ray tubes, sparking plug electrodes, instrument transformers where high magnetic permeability is required.

There are some recently developed nickel–iron–cobalt alloys whose outstanding characteristics are a constant low coefficient of thermal expansion, a constant modulus of elasticity and high strength. They are used in gas turbine casings, shafts and shrouds. Their low coefficient of expansion enables close control of clearance and accuracy in turbine shafts.

NICKEL–COPPER ALLOYS

In 1905 it was discovered that a Canadian nickel–copper–iron ore could be smelted to produce an alloy with good corrosion-resistance. The alloy, containing about 65 per cent nickel, 32 per cent copper and small amounts of iron and manganese, was trade-named 'Monel' after Ambrose Monell, the President of the International Nickel Company. It is resistant to sea-water corrosion and has been selected to protect the steel legs of some oil platforms in the vulnerable splash zones above and below the tidemark. It is also resistant to marine-growth such as seaweed and barnacles. A variation of Monel, containing aluminium, can be heat treated to give high strength as well as corrosion resistance and it is used for pump shafts and the propeller shafts of power boats. Monel is produced to this day, but by straightforward alloying.

COPPER-NICKEL ALLOYS

At the other end of the scale, the alloys rich in copper range from 90–10 to 60–40 compositions. The 90 per cent alloy features in ships' condensers and, in enormous quantities, in desalination plant for the Middle East. It has good resistance to corrosion by sea water and resists biofouling by marine organisms. A new use is as expanded mesh to contain fish in marine fish-farming units; the mesh keeps the predators out and the fish in, whilst the non-fouling copper–nickel keeps the mesh open to allow the food-containing sea water to flow through the enclosure. Another use under development is the cladding of ships' hulls to combat the fuel loss due to encumbrance caused by marine growth on the hull. The 70–30 copper–nickel–iron–manganese alloy is used for condenser tubes where environmental conditions such as sand-laden sea water exist.

Cupro-nickel alloys, in particular 25 per cent nickel, 75 per cent copper, are required for coinage; the present British 'silver' coins are made of this alloy, as are the American 5 cent piece or 'nickel' and the Indian one rupee. In addition to such familiar coins the Botswana one pula and 50 thebe, the Brunei 5 to 50 sen coins and the Zambian 50 ngwee are of cupro-nickel. The Libyan 10 dirham coin is of steel clad with cupro-nickel.

Nickel silvers are a series of copper–nickel–zinc alloys with varying amounts of nickel. The best quality E P N S – electroplated nickel silver – tableware is based on an alloy containing 20 per cent nickel. Although still in production, the demand for nickel silver has decreased as the popularity of stainless steel for tableware establishes itself.

17

MAGNESIUM

Magnesium has been called the lightweight champion among metals. Aluminium is more than one and a half times heavier, iron and steel are four times, and copper and nickel five times heavier. This low density, combined with a relatively high strength, has won for magnesium its present place in industry.

Although a comparatively late arrival in the world of engineering metallurgy, magnesium was isolated as long ago as 1808, earlier indeed than aluminium, and efforts to produce it commercially were made throughout the nineteenth century. The reasons for the tardy development of this metal, so lavishly provided by nature, are partly technical and partly economic. On the one hand extraction problems had to await the development of industrial electrolysis, and on the other hand there was little demand in industry for that lightness which is the metal's most characteristic property. Magnesium affords a good example of metal responding to the needs of man and fitting into the pattern of his civilization. It is significant that magnesium ores, unlike those of the other metals, are so widely distributed throughout the earth as to preclude the possibility of establishing a monopoly in them but the metal itself is so difficult to extract that it is exploitable only by highly industrialized communities.

Of all the common metals there is none about which there is so little public knowledge as magnesium. This is because magnesium is seldom employed for domestic articles, such as kitchenware and furniture, so we rarely have a chance to handle it. We come into frequent contact with iron, steel, aluminium, copper, zinc, lead, tin and nickel, and can usually distinguish one from the other, but if we see a piece of magnesium, we probably mistake it for aluminium. There is one important exception: the man in the street associates magnesium with the production of intense light, such as fireworks, incendiary bombs and military flares.

Consequently, most people have difficulty in believing that so flammable a metal can be used for any structural purpose, and they are quite incredulous when told that it has been employed for cooking-pots and frying-pans. The apparent contradiction is explained by the fact that before magnesium can be made to burn vigorously it must be melted, superheated and be in contact with ample supplies of air.

The special methods required for smelting and refining magnesium, casting the alloys, and for their later fabrication into various finished forms, were pioneered by the Germans, although a great deal of valuable development work was done in Britain and America. There has been a remarkable increase in output over the past eighty years. In 1900 annual world production amounted to only ten tons. This increased to 1000 tons in 1920, to 20 000 in 1937, and to a maximum of 238 500 in 1943 – an eleven-fold increase in six years, due to wartime needs of the aircraft industries. After the war, the annual tonnage fell to about 10 000, but after 1949 it rose rapidly and reached 215 000 tonnes in 1970 and about 400 000 tonnes in 1987.

Although pure magnesium has no great strength – about 100 N/mm^2 – suitably alloyed and worked its strength can be doubled or even trebled. The alloying elements most widely used are aluminium and zinc to increase the mechanical properties and manganese to improve corrosion-resistance. A typical casting alloy used for automobile engine components and chain-saws contains from 8 to 10 per cent aluminium, with 0·5 per cent zinc and 0·25 per cent manganese. At the other end of the scale lightweight beverage cans are made from an aluminium alloy containing about two per cent magnesium which provides stiffness. This use of magnesium in such an enormous industry accounts for about half the world's production of the metal.

In the search for alloy compositions with still greater strength, especially for aircraft, it was found that the addition of less than one per cent of zirconium has the property of reducing in size the individual grains of which the alloy is composed. For example the grain size of un-alloyed magnesium may be from 10 to 20 mm, a magnesium–aluminium alloy in the sand-cast condition has grains about 0·5 mm across, while a magnesium alloy with 0·65 per cent zirconium has a grain size of only about 0·03 mm. The smaller-grained metal is considerably stronger and more ductile. By the addition of other elements such as zinc, thorium, silver and rare earth metals, a number of alloys with great strength at room and elevated temperatures, creep resistance and good castability have been developed. Although the potential importance of zirconium as an alloying metal for magnesium was originally discovered by German metallurgists, the difficult alloying technique was perfected in Britain, and work on the new alloys proceeded simultaneously with the development of jet propulsion.

Further research work during the Second World War showed that thorium additions improved the resistance to creep of the magnesium–zinc–zirconium series of alloys, particularly in the cast state. The room-temperature mechanical properties are also generally superior to earlier alloys. Castings in magnesium alloys containing about 0·7 per cent zirconium, 3·0 per cent thorium, and 2·5 per cent zinc are used in both British and American jet engines. A still more important range of alloys contains about 0·7 per cent zirconium, 2 to 5 per cent zinc and 2 to 3 per cent cerium. A recent invention enables strong and ductile castings, for example the thrust reversers in the RB211 jet engine, to be made by heat-treating these alloys in hydrogen, which modifies their microstructure. This is an unusual use of hydrogen, since in most metals it reduces mechanical properties.

Magnesium alloys can be subjected to all the usual metallurgical treatments, including casting, diecasting, rolling, forging, extruding and pressing. They can be cast with ease and safety if precautions are taken to prevent the metal from overheating. Magnesium-based alloys are used mainly in the cast form.

No metal can be cut, drilled or shaped so easily or so fast as magnesium. This of course helps to reduce the cost of the final product. Magnesium also lends itself readily to welding and riveting. The strength of its alloys in relation to their weight is very high indeed. The cast alloys can show strengths of 300 newtons per sq. mm and wrought alloys up to 380 newtons. The strength-to-weight ratio of such materials resembles that of a high tensile steel.

The principal advantage of magnesium alloys is their lightness; they are widely used in chain-saws, textile machinery, automobiles and aircraft. During the war retractable undercarriage parts were cast in magnesium, pilots' seats were made of welded magnesium tubes, petrol tanks of magnesium sheet, while a large number of engine parts, such as supercharger blower casings, were also produced from magnesium alloys. In Britain alone nearly a million aircraft wheels were made of magnesium alloys.

The predominant use of magnesium in automobiles has been by Volkswagen. Up to 1980, 20 million of the famous 'Beetle' cars had been manufactured containing about 400 000 tonnes of magnesium alloy diecastings, the air-cooled engines being ideally suited for ultra-light alloys. The trend to produce front-wheel drive water-cooled engines reduced Volkswagen's consumption of magnesium, but the total tonnage is still very large and the plants in Germany, Brazil and Mexico contain impressive examples of magnesium diecasting technology. Several alloys are cast, containing aluminium from 2 to 9·5 per cent, plus small additions of zinc, silicon and manganese. The components include crankcases in the remaining air-cooled engines but the largest consumption of magnesium is represented by transmission housings, ranging in weight from 7 to 12 kg.

The light weight of magnesium alloys is a great advantage in chain-saws, wheels of sports-cars and aircraft, textile machinery, cameras, office and military equipment. Sporting uses include fishing reels, tennis rackets, archery bows and 'Little League' baseball bats.

It is interesting that, excluding the large requirements of Volkswagen, Germany still uses a greater tonnage of magnesium than any other European country. Once the metal is accepted as a viable material its use spreads, and in industries which produce chain-saws, for example, more and more components are made of magnesium. There are tremendous possibilities in the motor trade because weight reductions cause economies in fuel consumption. In North America automobile components made from magnesium alloys are gaining in popularity. By 1989 over 10 000 tonnes of diecast parts will be required.

Potentially, magnesium could replace aluminium in many applications, the choice between the two metals depending mainly on commercial considerations. However, the relative prices of the two materials are not the only factors. The machining of magnesium is much easier and faster than aluminium; as a further advantage magnesium does not attack die steels in the same way as aluminium, so when the alloys are diecast, longer die lives can be expected. This is benefiting the American producers because their automobile components are needed in quantities of many hundreds of thousands and the extension of die life is an added bonus.

Although some new high-purity casting alloys are more resistant to salt water corrosion than some aluminium alloys, as a general rule one has to allow for the extra cost of a protective chemical finish, to improve corrosion-resistance. Magnesium is a chemically reactive metal, though it does not corrode as much as iron and steel. To prevent deterioration, magnesium and its alloys can be protected by immersing the pieces to be treated in vats containing a hot solution of chromate salts. Black or golden oxide films, containing chromic oxide, are thereby deposited on the surface of the magnesium and these form a protective layer which acts as a base for the application of paint for further protection.

Large quantities of pure magnesium are required for alloying with other metals, as a deoxidizer or refiner in melting metals such as nickel, and in powder form in fireworks. Another use of magnesium is in the treatment of cast iron to nodularize the graphite, which makes the iron stronger and more ductile (see also page 144).

On page 232 reference is made to the 'sacrificial protection' of iron by zinc. For some applications magnesium is more effective, and it is employed extensively in America to protect oil pipelines and steel water tanks by connecting them to magnesium anodes. Another electrochemical property of magnesium is used for photo-engraving; indeed this method has already been used for book printing.

The raw materials for magnesium production are widely distributed. Each cubic kilometre of sea water contains over a million tonnes of magnesium compounds. Brines contain magnesium chloride; about two per cent of the weight of the earth's crust is of magnesium in the form of minerals such as dolomite or magnesite. Several processes have been developed to produce magnesium from dolomite, sea water and brine but all of them consume a large amount of energy.

Nowadays two electrochemical processes are employed. In one system crushed dolomite is roasted, then mixed with sea water and led to large tanks, where the insoluble magnesium hydroxide that has been formed settles to the bottom. It is then heated, to form magnesium oxide, which is mixed with coke and reacted with chlorine in large shaft furnaces, producing molten magnesium chloride. This is electrolysed with a high voltage, low amperage current. Chlorine is given off at the carbon anodes and recycled back to the process, while the liberated magnesium floats to the surface. Thus the magnesium is derived partly from dolomite and partly from sea water.

There is a newer process which is more energy-efficient. Brine, containing about 10 per cent magnesium chloride, is purified, concentrated and then the water is removed in several stages, first by evaporation, then by hot air and hydrochloric acid gas. The magnesium chloride is then electrolysed into magnesium and chlorine. This process is tending to supplant the dolomite-sea water process because it requires less energy and it yields the valuable by-product chlorine.

18

TITANIUM

As long ago as 1791 the existence of titanium was deduced by William Gregor, an English clergyman and mineralogist. He obtained only compounds of the element, not the metal itself, which was not isolated till over a hundred years afterwards. During the past four decades its industrial development has been spectacular. Only three tonnes were made in 1948 and by the late-nineteen-eighties free world production of wrought titanium had increased to about 30 000 tonnes per annum.

Titanium ores are widely distributed; of the structural metals only aluminium, iron and magnesium are more abundant. The processes of extraction, melting and fabrication are all expensive because of the need to avoid contamination, mainly by oxygen and nitrogen, while the metal is molten. Titanium ores such as rutile, containing titanium dioxide, are first concentrated and then converted to liquid titanium tetrachloride, which is roasted for several days in an atmosphere of argon, with additions of either sodium or magnesium metal, producing titanium and readily removable sodium chloride or magnesium chloride. The spongy titanium metal produced by this process has a purity of at least 99 per cent.

The extreme reactivity of titanium in the molten state makes it impossible to melt in normal crucibles because no refractory will resist it. Consequently a radically different melting technique had to be devised. The consumable-electrode vacuum-arc furnaces shown in *Plate 17* use a direct current arc between an electrode made of the metal to be melted and the base of a copper crucible cooled by water or by a liquid sodium-potassium alloy. The entire arrangement is enclosed in a vessel which either contains a non-reactive gas such as argon or which may be completely evacuated. The electrode is usually made of compressed blocks of titanium sponge, which is the condition of the metal obtained from the ore; alloying additions are usually included at this stage. Normally a slender ingot is melted first to act as the electrode for subsequent melting operations. In spite of their reactivity when molten, titanium and its alloys are cast directly into finished shapes for aerospace and many other uses.

About half the annual production of metallic titanium is represented by the commercially pure metal, which is strong and ductile. The remainder is required as alloys. The combination of high strength, lightness and corrosion-resistance accounts for the value of titanium and its alloys in modern technology. Its strength-to-weight ratio provides weight savings of at least 40 per cent when it replaces steel. Although highly reactive when molten, the solid metal is resistant to a wider range of corrosive substances than stainless steel.

The alloys of titanium divide into several groups, depending on the amounts of alloying elements and the resulting crystal structures. Current alloys contain up to 25 per cent of added elements such as aluminium, vanadium, molybdenum, niobium, zirconium and tin. These alloys are readily fabricated into complex, high-integrity parts for aircraft structures. One such alloy, containing 6 per cent aluminium and 4 per cent vanadium, is used world-wide in gas turbine engines and airframes. Recent development work has revealed that this and similar alloys exhibit superplasticity (the ability to be stretched several hundred per cent without breaking) at low strain rates and at temperatures around 900°C. The new technology of superplastic forming is now applied routinely, primarily for airframe components. In many instances parts are joined during the same forming process by diffusion bonding, whereby the titanium alloy surfaces weld together by atomic diffusion, leaving virtually no trace of a joint-line.

Alloys have now been developed for rotating parts in aero-engines, capable of operating up to 600°C. Some titanium alloys can be heat treated to strengths of over 1400 N/mm², a magnificent combination of strength with lightness; they are used where maximum strength per unit weight and fatigue resistance are essential.

Concorde, flying at Mach 2, includes about 4 per cent of titanium alloys, mainly in the engine surrounds. The European Airbus family uses a wide variety of titanium products, including formed and welded firewalls, wing access panels, landing gear brackets and many other components. Similarly a number of modern helicopters rely on titanium alloys for rotor heads, blade attachments, fuselage beams, anti-vibration mountings and weapon carriers. Current designs of civil aircraft contain up to 10 per cent of titanium alloys, while military aircraft such as the Tornado and Jaguar have up to about 26 per cent. The Lockheed Blackbird, capable of over Mach 3, is said to contain about 85 per cent of its structural weight in titanium alloys.

Plate 18 shows one of the RB211 series turbofans, which has titanium alloy compressor discs, spacers, rings, casings and blades, some of which are the recently developed hollow 'widechord' blades. In the U.S.A. titanium alloys are used for airborne military equipment, such as mortar bases and armour plating. Outside the aerospace industry, titanium is being specified for many engineering purposes that call for a high strength-to-weight ratio. Typical applications include steam-turbine blading, connecting rods, crankshafts, camshafts, outlet valves and springs in high performance engines.

Titanium is used in the production of chlorine, because it is one of the few metals that are not corroded by the gas. Other functions in the chemical industry include the production of acetic acid, benzoic acid and ethylene amines. Recent developments of titanium to combat 'condenseritis' will be discussed on page 234. Because of its complete inertness to body fluids titanium is now widely used for surgical implants.

In 1545 the warship *Mary Rose*, waiting to engage the French fleet, sank suddenly with the tragic loss of seven hundred men. Over four centuries later, the waterlogged remains of the ship were lifted to the surface and brought back to Portsmouth. Titanium was chosen to support the deck beams because of its light weight and resistance to corrosion. Forty-five adjustable vertical props and several horizontal ones were used. A constant spray of chilled water plays over the structure of the ship to prevent premature drying-out of the delicate timber, which will be sprayed with polyethylene glycol, during the next 10–15 years. Finally the structure will be dried under controlled conditions.

The life of high-speed steel cutting tools is improved by coating them with a deposit of titanium carbide, or by depositing an underlayer of titanium carbide, followed by a series of gradations of titanium carbonitrides and finally a titanium nitride layer. This gradual change in composition is necessary to maintain adhesion of the hard titanium nitride layer. Recent developments incorporate the alternate deposition of titanium nitride and alumina, totalling ten layers, each about 0·5 micron thick. Such products have increased tool life by as much as six times.

It will have been gathered from the chart on page 13 that the amount of titanium ore that is mined is much greater than can be accounted for by the production of the metal and its alloys. By far the largest use of titanium is in a non-metallic form, as titanium oxide for paints, where it has a greater whitening effect than lithopone or white lead carbonate, because of its permanence and its ability to reflect white light.

19

SOME MINOR METALS
and a few other elements used in metallurgy

ANTIMONY. In appearance this semi-metal is similar to zinc. Being hard and brittle (it can be powdered with a hammer), it is rarely used alone, but is employed as a hardening addition to lead, solder, pewter, white metal bearings, accumulator plates, telephone sheathing and in shotgun ammunition. Considerable quantities of antimony trioxide are used in flame-retardant compositions

ARSENIC. Like antimony, this element is not a true metal, though it has some metallic properties. During the whole history of metallurgy, arsenic ores have been found in conjunction with those of copper, so some arsenic was contained in the copper when it was smelted. It was found to be advantageous, in conferring increased hardness when the copper was cold-worked. Arsenic is alloyed with lead for sporting ammunition shot and, in conjunction with antimony, for lead cable sheathing. Arsenic, as well as boron and phosphorus, is used as a dopant in the manufacture of silicon integrated circuits.

BARIUM. This silvery-white metal is spontaneously flammable in moist air and, though valuable in the form of some of its compounds, has only a few direct uses in metallurgy; for example it has been included, in association with calcium, in bearing metals manufactured in Germany. Compounds known as ferrites (not to be confused with ferrite, discussed on page 133) are manufactured for use as magnetic components of electrical equipment. They contain the iron oxide Fe_2O_3 combined with the oxide of barium, or other oxides such as those of strontium or lead. They are magnetic but are also electrical insulators. Ferrites have been used in the memory circuits of computers, though recently they have been superseded by silicon.

BERYLLIUM. Beryllium is lighter than aluminium and nearly as light as magnesium; it has a melting point of 1279°C – much higher than that of aluminium. It has good corrosion-resistance in air at temperatures below 500°C, also a high strength-to-weight ratio. Beryllium was considered to be a wonder metal with a promising industrial future for the construction of aircraft fuselages and possibly for missile bodies and space vehicles. Unfortunately, difficulties were encountered in processing the metal to develop it into suitably workable material; consequently interest in industrial application has developed more slowly than had been prophesied.

The raw material has to be converted to a metal ingot by vacuum-melting in an electric induction furnace. Beryllium is then machined down to a fine powder, which is heated and compacted in a vacuum, into the required shape of rods, bars or tubes. The handling and fabrication are made additionally complex because beryllium has toxic properties.

The most important alloys containing beryllium are those with 0·5 to 2·0 per cent of the metal, with copper as the main constituent. An alloy of copper with 1·8 per cent beryllium which has been rolled, heated to 800°C and quenched in water has a strength as great as mild steel. If this quenched alloy is reheated to 335°C for two hours and then cooled, precipitation hardening (see page 163) takes place and the alloy develops a strength two or three times greater than that of mild steel. This 'beryllium copper' is among the strongest of non-ferrous alloys.

BISMUTH. Apart from medicinal compounds of this metal, it has a few uses in metallurgy; for example it is a constituent of fusible alloys, which were mentioned on page 44. In making malleable iron, small amounts of bismuth are added in pellet form to the contents of the ladle before pouring, to help stabilize the carbide which is an essential feature of the malleable iron. The compound manganese bismuthide is magnetic and has attracted interest when produced in the form of thin films for advanced computer memories.

BORON. Principally this non-metal is used in the form of borax (sodium borate combined with molecules of water) and boric acid, and is required for the manufacture of porcelain, enamels and glass, in dyeing textiles, fire-proofing timber, as a food preservative and for pharmaceutical purposes. Boron has a great capacity to absorb neutrons, so steel containing boron has been used in nuclear power stations for control rods (see page 271).

In ferrous metallurgy small quantities of ferro-boron are added to improve 'hardenability' (the degree to which steel will harden when quenched). Great hardness is a quality associated with the metallurgical uses of boron, and relatively inexpensive steels may be 'boronized' to give surfaces more hard-wearing than those produced by carburizing or nitriding. In non-ferrous metallurgy boron and titanium are added to refine the grain size of cast aluminium alloys.

Compounds formed between boron and titanium or zirconium are practically as hard as diamond; they are better conductors of electricity than some pure metals and have very high melting points of around 3000°C. They are resistant to corrosion by molten metals and salts and, now that they are becoming commercially available, these borides are expected to find many new applications.

CADMIUM. Zinc ores contain on average 1 part of cadmium to 200 of zinc and the metal is produced as a by-product. Its main use is for cadmium plating steel components, especially those to be soldered or requiring lubricity. Nickel–cadmium rechargeable batteries are used in standby and emergency power stations and in the starting of heavy diesel engines. Cadmium alloys are used as solder for aluminium, for bearings, and as low temperature fusible alloys. The use of cadmium to strengthen copper in overhead electric cables is described on page 172. The yellow-red compound cadmium–sulphide–selenide colours many plastic products, paints and ceramic glazes.

CAESIUM. This metal possesses the highest thermal expansion – about eight times that of iron and steel. Apart from mercury, caesium has the lowest melting point of all metals: 28·6C. It is used in 'atomic clocks' and features in extremely sensitive light detectors and instruments which measure time intervals much less than a millionth of a second

CALCIUM. Some of calcium's chemical compounds are well known – for example, limestone and chalk are calcium carbonate and the 'chalk' for writing on blackboards is calcium sulphate. Calcium is a soft metal and corrodes rapidly in air so it is not used on its own. A lead alloy containing one per cent of tin and 0·08 per cent calcium is attracting interest, particularly in the U.S.A., for so-called 'maintenance-free' car battery plates; this refers to the fact that the battery does not need water additions in the course of its normal life. The importance of metallic calcium has increased since it became used in the process of winning

CHROMIUM. The electroplated layer of chromium, only about 0·0003 mm thick, that is deposited over nickel plate on steel, brass or zinc alloy is familiar to everyone. The most important use of chromium, however, is in alloy steels, where it confers increased hardness after heat treatment. In amounts of up to 12 per cent, it improves the resistance to scaling or oxidation of iron at high temperatures. There are several low and medium alloy steels containing chromium with other elements, such as molybdenum and vanadium, which can be used continuously at temperatures from 500°C upwards to 700°C, for components of power stations and steam turbines. Additions of over 12 per cent chromium give the range of stainless steels. When required as a strengthening and hardening addition to cast iron or steel, the chromium is provided as ferro-chromium, containing 70 per cent chromium, 4 to 6 per cent carbon and the rest iron. Ferro-chromium has a lower melting point than that of the pure metal and is a cheaper and more convenient way of adding it. Resistance wires for electric heating elements contain 80 per cent nickel and 20 per cent chromium. This range of alloys was the base from which the 'Nimonic' alloys were developed for jet aero-engines, discussed on page 193.

COBALT. This metal has a wide and important range of applications but practically all of them involve cobalt as an alloying element or as a chemical compound. The largest use is in the field of superalloys which were discussed on page 193. Although most of these, such as the Nimonics, contain only a small percentage of cobalt, the total tonnage amounts to about 30 per cent of the requirement for the metal. Next in importance come the various new magnetic alloys. Until recently the alloy with 35 per cent cobalt, 57 per cent iron, 5 per cent tungsten, 2 per cent chromium and 0·9 per cent carbon was the most highly magnetic alloy that had been discovered. Since then, there have been outstanding improvements thanks to the association of cobalt with various of the rare earth metals listed on page 218. Among these, a cobalt–samarium alloy and one with cobalt and praseodymium are the most magnetic materials known at present, but undoubtedly there will be fresh fields to conquer as all possible combinations of cobalt with the fifteen rare earth metals are fully explored.

A range of cobalt-rich alloys known as Stellite are extremely hard and resistant to corrosion and they retain these properties at elevated temperatures. A typical alloy contains 26 per cent chromium, 5 per cent tungsten, 1 per cent carbon and 68 per cent cobalt. Another has 27 per cent chromium, 2 per cent nickel, 6 per cent molybdenum and the balance cobalt. Such alloys can be cast, to make gas turbine blades and extrusion dies. The Stellite alloys can also be applied by welding, to form a very hard and abrasion-resisting deposit on the valves for steam and chemical plant. A small but, to many sufferers, important use of Stellite alloys is in the rebuilding of knee and hip joints in the operations now performed successfully by surgeons.

Other cobalt-rich alloys, containing 34 per cent chromium, 19 per cent tungsten and 2 per cent carbon, are formed into cutting-tool tips and milling cutter blades. The metal is used as the bonding agent for tungsten carbide in tipped tools.

It will be seen, therefore, that cobalt plays an important part in several modern technologies. Unfortunately, however, well over half the free world's supplies come from Zaire, one of the African countries that has had its full share of troubles. Consequently the price of cobalt has fluctuated dramatically. At the beginning of 1977 the metal cost about £7000 per tonne but by the end of 1978 it had soared to £21 000 per tonne. In view of the high price, other countries with workable ore deposits began to extract the metal, thus increasing supplies but in spite of these efforts, price movements remained erratic. In early 1982 the free market price was £12 000 per tonne; if fell to £9000 in 1983, rose to £17 000 in mid-1984 and in early 1989 was about £11 000. Of course cobalt occurs as a by-product in, for example, the smelting of nickel, but the quantities are not large and they could only be increased by artificially expanding the output of nickel, which would cause un-necessary over-production.

GALLIUM. A soft, silvery-white metal, gallium is unusual in having a wide range between its melting point, about 30°C, and its boiling point, about 2070°C. Gallium is one of the elements that have been used in 'doping' silicon for transistors. The compound gallium phosphide features in light-emitting diodes, tiny electro-luminescent 'lamps' which are used in clusters to form digital read-out displays. The recent development in miniature gallium arsenide lasers is leading to a revolution in telecommunication technology.

GERMANIUM. This semi-conductor element was one of the first materials used for transistors, which are amplifiers of small size and great reliability using only a millionth of the power required by the thermionic valves which they replaced. Transistors were developed as a result of research in the nineteen-forties, in which semi-conductor materials were 'doped' with minute amounts of other elements, producing an enormous change in their electrical properties. Electrodes were attached to a single crystal of germanium, which was doped with arsenic at one electrode and with gallium or indium at another. This 'diode' has the property of different electrical conductivities in opposite directions. By fixing two diodes together, back to back, a transistor is formed. As will be seen when we discuss silicon, that material has replaced germanium for most transistors. However, germanium is sometimes preferred for specialized applications such as infra-red detectors.

GOLD. To produce one troy ounce (31·3 g) of gold from a typical rich mine, about three tonnes of ore must be blasted from the reefs, conveyed to the shaft, hoisted to the surface, crushed to the consistency of face powder, then passed through the processes of agitation, filtration and treatment with potassium cyanide to separate the metal. The production of this small amount of gold also requires over 5000 litres of water, 750 kWh of electric power, explosives, chemicals, compressed air and 39 man-hours. *Plate 2* shows one of the world's largest gold mines and there are some notes about mines on page 15.

Gold has complete resistance to atmospheric attack and oxidation, as was shown when the tomb of Queen Pu-Abi was excavated in Meso-potamia (now Iraq), 4600 years after the death of the lady, and when Tutankhamun's tomb was discovered by Howard Carter after 3300 years. The most malleable and ductile of all metals, gold can be beaten to leaf so thin that it is transparent. In a number of civil and military aircraft a film of gold approximately 0·000 005 mm in thickness, in a sandwich of laminated glass,* is used as an electro-conductive system to provide anti-ice and anti-mist capability. The vital electronic systems that enabled man to take his first step on the moon were protected from take-off blast by coatings of gold, and it was used in the circuitry of computers on the moon voyage.

For many uses of gold the pure metal would be too soft on its own and it is therefore hardened by the addition of alloying elements, copper being the most common; in addition, silver, nickel, palladium and zinc are included in jewellery. Copper alone produces reddish alloys while with silver alone the colour becomes greenish-white. Together they produce the familiar rich yellow colour; for lower qualities zinc is added.

Rolled gold consists of a thin layer of gold alloy, usually 12 or 14 carat, bonded on a base metal such as brass or nickel silver. The process involves high temperature and pressure and the final product has a

---

* Information about a new development, involving indium, is given on pages 210–11.

perfect adhesion between the two metals, matching the appearance of the gold but without its high cost.

The gold content of jewellery alloys is designated by the carat system. The term refers to a 24th part – thus 18 carat contains 18 parts of gold and 6 parts of alloying elements. The compositions used in the U.K. for jewellery are 22, 18, 14 and 9 carat. (12 carat gold is used for rolled gold.) These are legal standards and gold offered for sale must be submitted to one of the four Assay Offices for checking the gold content and hallmarking to one or other of these standards. Hallmarking is the oldest consumer protection system in the world, having been carried out by the Worshipful Company of Goldsmiths in the City of London since A.D. 1300. Sometimes the gold content is expressed as a decimal; thus 9 carat is 0·375.

The possession of gold is a sign of wealth, whether in the safe of a collector, in a fine display of gold teeth or in the several hundred thousand ingots stacked at Fort Knox. Although sovereigns are no longer used as currency in Britain, the Royal Mint still makes them in considerable quantity – they are exported for purchase by collectors or for use as an international currency in Middle and Far Eastern countries.

The standard ingot, measuring $175 \times 90 \times 40$ mm, weighs 400 troy, or 439 avoirdupois ounces (12·45 kg). Price variations depend on supply, demand and political events. When Afghanistan was invaded in December 1979, the price of gold escalated from 600 to 850 dollars per ounce, and a standard ingot would have cost 340 000 dollars. In early 1989 a combination of large production and reduced demand caused the price to fall to under 400 dollars per ounce: the ingot cost a mere 160 000 dollars.

HAFNIUM. Zirconium ores contain compounds of hafnium in very small quantities; the two elements are chemically similar and at first it was difficult to separate them; the need to do this is explained on page 227. However, recent developments of the solvent extraction process, discussed on page 18, have made it much easier to separate the hafnium from zirconium. It is a very effective metal for absorbing neutrons, making it specially suitable for some nuclear reactor control rods, where expense is no object.

INDIUM. In 1928 the world output of this metal was only one gram. Though it is now produced in some quantity indium still ranks as rare and precious. Ultra-high-purity grades of the metal are being developed for use in semi-conductors. Indium is used for the surface coatings of some bearings, to give improved resistance to corrosion and wear. A three-layer bearing is produced; the back, of low-carbon steel, is lined with a copper–lead–tin alloy which, after broaching or boring, is plated with a layer of pure lead and machined accurately to size; a thin layer of indium is then electronically deposited on the lead. The bearing is heated and the indium diffuses through the bearing, almost to the interface between the steel back and the lining.

The function of indium is threefold. Firstly it reduces the susceptibility of the copper–lead layer to the corrosive attack of lubricating oils; secondly it provides an increased fatigue strength to the lead overlay; and thirdly it provides good anti-friction properties. The use of indium-faced bearings has extended to medium-heavy diesel engines and in the coming years bearings larger than 200 mm, the present manufacturing limit, will be achieved.

In the search for improved anti-ice and anit-mist materials for laminated aircraft windscreens, Triplex have treated an alloy of indium with tin by sputtering. The composite oxide so formed is deposited within the laminate to a thickness of about 0·0025 mm, that is 250 times thicker than the gold which was used previously, but this deposit is transparent and of a more neutral colour than the gold film.

IRIDIUM. Having a specific gravity of 22·56, iridium, a member of the platinum group of metals, is second only to osmium in heaviness. Iridium's corrosion-resistance is even better than that of platinum, but despite a very high melting point of 2454°C, it is less oxidation-resistant than either platinum or rhodium. Iridium is difficult to work and only recently has begun to be used in its elemental form. Iridium-tipped sparking plugs have been developed to replace platinum-tipped plugs in aero-engines. Platinum alloys containing up to 30 per cent iridium are used for jewellery; the 10 per cent alloy is used for diamond settings in the U.S.A. Platinum–iridium alloys are used for standards of weight and length because of their great stability and permanence.

After exploring the atmosphere of Jupiter and its moon Io, Voyager 1 left the solar system. Its sister, Voyager 2, reached Uranus in 1986 and is due to approach Neptune and its moon, Triton, in 1989. These car-sized space vehicles were provided with heat sources consisting of three sets of eight ceramic spheres of plutonium dioxide encapsulated in thin iridium shells, with special vents to release the gas products of the radioactive decay.

LITHIUM. This metal has only half the density of water. It reacts vigorously with air and water, but in spite of such problems the new aluminium–lithium alloys discussed on page 281 are being made success-fully and used in aeronautics. The metal is unusual in the considerable range between its melting point of 179°C and its boiling point, 1336°C. This may be compared with its sister metal sodium, which melts at 97°C and boils at 880°C.

The low density and small atomic weight of lithium, combined with its high reactivity, make it an attractive material for the negative electrodes of batteries that are required to provide a lot of power for their weight. Efforts to overcome the formidable practical problems have been continuing for the past ten years. The lithium is often used in conjunction with very reactive, toxic, active materials such as thionyl chloride and sulphur dioxide; the electrolytes consist of lithium salts dissolved in non-aqueous solvents. 'Pacemaker' cells are manufactured in which lithium is opposed by silver chromate positive electrodes.

Two isotopes of lithium are significant in nuclear engineering. Lithium-7 does not absorb neutrons in nuclear reactors, but lithium-6 has a great ability to absorb neutrons and hydrogen isotopes are derived from it. Perhaps in the twenty-first century mankind's energy problems will have been solved, thanks to the 'taming' of thermo-nuclear fusion, for which experimental equipment is now being built. Lithium's properties of low melting point and high boiling point referred to above have already made the metal suitable as a heat transfer agent in some nuclear reactors and it may be chosen to extract and transfer the colossal heat from the thermo-nuclear reactors of the future.

MANGANESE. Each year about ten million tonnes of manganese ore are mined, containing an average of 35 per cent metal. It is therefore fifth in tonnage, after iron, aluminium, copper and zinc. The metal is never used on its own but usually extracted in the form of ferro-alloys; one of these contains 80 per cent manganese and another has about 20 per cent. A third alloy has about 20 per cent manganese and 10 per cent silicon. Such ferro-alloys are added to steel during the course of manufacture. On its own manganese is brittle and useless for engineering but, when alloyed in steels, it helps to combat brittleness and it increases toughness.

One important feature of manganese is that it reduces the deleterious effect of sulphur in steel. Iron sulphide forms continuous films which cause the metal to become embrittled. When manganese is present it combines with the sulphur to form manganese–iron sulphide, which is distributed in small particles and not harmful. Most low-alloy steels contain from 0·2 to 1·2 per cent manganese to provide toughness. With 1·3 to 1·8 per cent the strength of steels is further increased; such materials are used in automobiles, where the greater strength permits weight reduction. Steel for bridges and ship plates contains 1·2 to 1·6 per cent manganese, with 0·25 to 0·30 per cent carbon.

Very hard steels with 10 to 14 per cent manganese have been mentioned on page 148. They are used, in the water-quenched condition, where great resistance to abrasion is necessary: in railway points and crossings, rock drills and stone crushers.

Manganese bronze is a high-grade brass with 40 per cent zinc containing manganese which toughens and strengthens the alloy. The metal is an important constituent in alloys of aluminium, nickel, silver, titanium and copper-base electrical resistance alloys and corrosion-resistant alloys.

MERCURY. The only metal that is liquid at ordinary temperatures is mercury or 'quicksilver'. It is used in making the chemical compound mercury fulminate ($Hg(ONC)_2$), which is one of the 'initiators' used to detonate explosives. Mercury also appears in various scientific instruments such as thermometers, barometers, discharge lamps, vacuum pumps, and in many forms of small electric contact-breakers. Metals such as tin, silver, and gold dissolve in liquid mercury to form alloys or 'amalgams' which are used in dentistry.

MOLYBDENUM (commonly called 'Molly'). Though molybdenum was first isolated in 1799, it has only recently come into the ranks of important minor metals. It has a high melting point: 2625°C, which is exceeded only by tungsten, rhenium, tantalum and osmium. By far the largest ore deposits and metal production are in the western hemisphere, with the U.S.A. contributing the biggest share, followed by Canada and Chile. The largest mine is at Climax, Colorado, at a height of 3500 metres; more than 500 kg of ore must be mined underground to recover one kilogram of molybdenum metal.

The steel industry absorbs about 80 per cent of the total molybdenum production, mostly in low alloy steels for constructional and other uses. When added to alloy steels containing nickel and chromium, molybdenum reduces the tendency to embrittlement that would otherwise occur when such steels are heat-treated. Among many other new developments, molybdenum has played a part in improving the efficiency of the rail traffic hauling iron ore. In one Australian mine there are ten unit trains each consisting of about 140 cars holding about 100 tonnes of ore, travelling at speeds up to 50 kmph. Ordinary carbon steel rails were wearing out rapidly under such massive and continuous loads;

chromium–molybdenum alloy steel rails have replaced the carbon steel ones, with significantly better results.

About 20 per cent of all molybdenum production goes as small additions, up to about 3 per cent, to stainless steels. In another related use nickel–chromium–molybdenum steels are used for the pipelines that convey oil and gas from Alaska, North Canada and Siberia, where temperatures are as low as minus 60°C. About 10 per cent of molybdenum production goes to the making of tool steels. Research work in the U.S.A. led to molybdenum replacing tungsten in high-speed steels; it is generally reckoned that molybdenum has twice the 'power' of tungsten, so that a steel previously containing 18 per cent tungsten could be replaced by one with 9 per cent molybdenum. In practice chromium and vanadium are also added to high-speed steels. The inclusion of the metal in H.S.L.A. and dual-phase steels has already been mentioned on page 151.

There are numerous small but important uses of the metal such as for radiation-shielding in vacuum melting furnaces and as a construction material in nuclear engineering. A compound of molybdenum has important uses in the field of lubrication: molybdenum disulphide is an effective addition to industrial and automobile lubricants which have to withstand high pressures and temperatures.

NIOBIUM. The largest consumption of this metal, known to the Americans as columbium, is indirect – as ferro-columbium additions in steel-making. Like most 'new' metals niobium has been known for a considerable time but a full appreciation of its potential value became apparent only when major advances in extraction and fabrication techniques enabled metallurgists to provide the metal for its most exciting uses. Niobium has a high melting point of 2468°C and several attempts have been made to develop niobium alloys suitable for components of spacecraft, so far without much success. However it is resistant to attack by liquid metals, so niobium components are used in sodium vapour lamps for street lighting.

Niobium does not absorb neutrons readily, thus making it suitable for fuel cans and support brackets in nuclear reactors. The resistance of pure niobium to the attack of liquid sodium leads to its use for fuel canning material in liquid-metal-cooled reactors.

Some niobium alloys and compounds become superconductive at very low temperatures, typically below −263°C (10K). They lose all resistance to electric current. The materials include a niobium alloy with 44 per cent titanium, in the form of fine filaments in a copper matrix, or the niobium–tin intermetallic compound Nb_3Sn, as a tape with copper backing. During the past few years dramatic discoveries have been made in the field of superconductivity and these will be discussed on page 289.

OSMIUM. The main interest of this rare metal lies in its neck-and-neck contest with iridium to be the heaviest of all elements. When a small porosity-free piece of osmium made by arc melting in a vacuum is tested, its specific gravity is 22·59 grams per cubic centimetre. A piece of iridium sheet has been measured with a specific gravity as high as 22·56. On this count osmium wins. However, when the theoretical specific gravities of the two metals are calculated from the lattice spacings, osmium would have a specific gravity of 22·60 and iridium 22·65. Until a piece of iridium has actually been produced heavier than osmium, we, and no doubt the *Guinness Book of Records*, will award the trophy to osmium. In its elemental form osmium is virtually useless, as it forms an oxide which volatilizes at room temperature. The metal got its name from the Greek word for 'a smell' because its oxide possesses a peculiar odour.

PALLADIUM. A member of the platinum family of metals, palladium is the one most similar to platinum, though it is slightly less corrosion-resistant and has a lower melting point. Palladium is considerably cheaper than platinum, has a lower density and can replace it in some applications where conditions are not extreme, including electrical contacts in speech circuits of telephone systems. Palladium catalysts are used in oxidation and hydrogenation processes, in margarine manufacture and the production of ethylene from acetylene. This metal has the unique property that it can rapidly absorb up to 900 times its own volume of hydrogen. Super-purity hydrogen for chemical and metallurgical processes is obtained by passing impure hydrogen through a diaphragm of heated palladium which acts as a filter, excluding other gases. In recent years a series of palladium-containing brazing alloys has been developed. They are particularly suitable for joining stainless steel and cobalt alloys, as well as the nickel–chromium high temperature alloys.

PLATINUM. Originally this was found as a native metal in alluvial deposits in Colombia and the Yukon, but today the only important supplies outside Siberia are from South Africa, where it is recovered from a sulphide ore, and Canada, where it is found associated with the Sudbury nickel/copper ores. Platinum has exceptional resistance to corrosion, being superior even to gold in this respect; it has a high melting point of 1773°C. It therefore is applied where even the slightest amount of oxidation or corrosion would be detrimental – for example for electrical contacts carrying small voltages, and crucibles and other pieces of apparatus used in critical chemical analyses. For many years just a curiosity, platinum alloyed with copper, iridium or ruthenium eventually became popular as a jewellery alloy, particularly for the mounting of diamonds on rings. The metal is hallmarked with an orb.

Platinum catalysts are used in many chemical processes, the most important of which is the catalytic oxidation of ammonia to form nitric acid, with large woven gauzes of platinum alloyed with rhodium. In the U.K. 10 per cent rhodium is required but in the U.S.A., where the process is done at very high pressures, they use only 5 per cent of the expensive rhodium addition. A few years ago pure platinum was used as a catalyst in reforming low-octane petrol but now the metal is alloyed either with small amounts of rhenium or with silicon and germanium.

The efforts to reduce pollution and smog caused by hydrocarbons, carbon monoxide and oxides of nitrogen have involved the use of platinum catalysts in automobile feedback systems, which self-adjust according to the composition of the exhaust gases. The fuel must be lead-free, because leaded petrol poisons the effect of the catalyst. Platinum is alloyed with rhodium to improve the efficiency, often in the proportion five parts platinum to one part rhodium. It is also alloyed with palladium, and sometimes pure palladium is employed, the various materials being selected according to the requirements of the vehicle. Consequently the automobile industry has become the largest user of platinum. So far these catalysts have been required on a large scale in America, Japan and Sweden. The rest of the world, including Britain, has been slower than anticipated in demanding stringent emission control. Progress depends on two factors: first, the will to improve the environment and second, the increasing availability and sale of lead-free petrol, without which the catalysts do not function. The requirement for platinum might have resulted in a major shortage but for the exploitation of new, though leaner, deposits in South Africa. In the meantime, Ford has been developing systems that do not require platinum.

Optical glass is melted in large platinum-lined crucibles to avoid contamination by refractories; platinum-clad apparatus is used in many other glass-making processes. Continuous filament glass fibre is made by melting the glass in solid platinum–rhodium alloy vessels which are heated by electrical resistance, the glass filaments being drawn off through nipples formed in the bottom of the vessel. The most accurate methods of measuring high temperatures up to about 1400°C are either platinum resistance pyrometers or platinum/platinum–rhodium alloy thermocouples. These are widely used for the direct measurement of the metal temperature in steel making.

PLUTONIUM. In March 1941 this was first made synthetically by bombarding uranium atoms in the cyclotron. Such bombardment leads to the formation of neptunium, which is unstable and is then transformed into plutonium. During the operation of a nuclear reactor, plutonium is formed from uranium 238 (see page 257); it can be separated chemically from the uranium and is then available for fast reactors. Plutonium is fiendishly toxic, highly reactive and a bone-destroyer. It is more fissile than uranium 235 and is an alternative for it in atomic bombs. Because plutonium is a fissile material, it has a 'critical mass' at which the fission chain reaction becomes self-sustaining and the hazards of radiation and explosion become imminent. The critical mass depends on the geometry of the system; it is higher for a flat shape than for a sphere. It also depends on the nature of the material, whether the plutonium is in the form of metal, oxide or solution such as plutonium nitrate. The critical mass for plutonium is very much smaller when it is in solution than when it is in solid form.

POTASSIUM. This reactive, soft, silvery-white metal has chemical properties similar to those of lithium and sodium. At about 50/50 composition potassium and sodium form a eutectic melting at room temperature and this alloy is used for cooling in the consumable-arc titanium melting furnaces. Potassium compounds are widely used in chemical processes, including the extraction of gold by potassium cyanide.

RADIUM. This extremely rare, highly radioactive, heavy metal is extracted from uranium ores, where it occurs in the proportion of about one part radium to three million parts uranium. The fantastic scarcity of radium may be assessed from the production of Canada, the world's largest source of supply, which amounts to only a few kilograms per annum, accurate production figures being almost impossible to ascertain. The metal is chiefly used for medical purposes on account of the penetrating radiation that it emits continuously, which is used for the treatment of cancer and some skin diseases.

Very minute amounts of radium compounds are mixed with zinc

sulphide; the radiation causes the sulphide to become luminous and this is used for dials of watches, compasses, and also in luminous gun sights. This accounts for about a tenth of the output of radium. The historical interest, the medical value of radium, and the inspiration of the story of Marie Curie's endeavours are very great – but metallurgically radium has little significance.

'RARE EARTH' METALS. From time to time mention is made in technical journals of elements with tongue-twisting names, like praseodymium, which are arousing new interest. They represent about one seventh of all the known elements but are scarce, expensive and produced in only small quantities. They are known also as lanthanides, because lanthanum, with an atomic weight of about 139, is the first of the series; lutetium, with an atomic weight of about 175, is the last. They are all somewhat similar to aluminium in their properties. An alphabetical list of their names is as follows:

cerium	gadolinium	neodymium	terbium
dysprosium	holmium	praseodymium	thulium
erbium	lanthanum	promethium	ytterbium
europium	lutetium	samarium	

Scandinavian scientists played an important part in isolating these metals and the names given to several of them reflect this influence. The mineral gadolinite was named after the Finnish chemist Johann Gadolin, hence the name gadolinium, used for control rods for very special nuclear reactors. The mineral had been discovered near the Swedish town Ytterby and from that the names of ytterbium, erbium, and terbium were derived. Holmium was named from Stockholm, thulium from Thule, the early name for far-northern Europe.

The Russian mine officer Samarski lent his name to posterity with the metal samarium, which has become very important among the highly magnetic alloys. Lutetium was named from Lutetia, the ancient name for Paris, and europium derived its name from Europe. Cerium was named after the asteroid Ceres.

Some of these metals borrowed Greek roots for their names. Dysprosium means 'the one hard to get at'; lanthanum 'the hidden one'. The name didymium, meaning 'twin', was given to a material which was found to be composed of two elements which, when isolated, were called neodymium, the new twin, and praseodymium, the leek-green twin.

While they do not hold an important commercial position in the metal world, the rare earth metals may be expected to mark up small but notable achievements in the future. Potential fields for development include additions to some alloy steels, components in nuclear energy plant, solar energy devices and electronics. The main constituent of cigarette-lighter flints, in the form of 'misch-metall', is an alloy containing cerium and other rare earth elements. When this alloy is rubbed by the steel wheel of the lighter, a substantial spark is formed; the misch-metall is said to be pyrophoric.

The average household contains about sixty small permanent magnets in domestic appliances and audio-visual equipment. Some of these will be of samarium–cobalt alloy and if the equipment is more up to date, the magnets will be of the boron–iron–neodymium alloy which has the highest magnetic power so far discovered.

The family of rare earth metals has been termed a Pandora's box, but it has been pointed out that so far the list does not include either 'delirium' or 'pandemonium'.

RHENIUM. This is a very heavy, silvery-white metal with the second highest melting point, 3167°C, and considerable resistance to corrosion. It is used as a platinum–rhenium catalyst in the petroleum industry.

RHODIUM. When William Wollaston succeeded in separating a previously unknown metal from platinum in 1804, he decided to call it rhodium, because some of its compounds were coloured rose-red – the Greek word for rose is *rhodon*. The metal has similar properties to those of platinum, including great resistance to corrosion, but it is lighter than platinum and has a higher melting point. Its principal use is as an alloying addition to platinum, for example to make the spinnerettes required in the manufacture of glass fibre and synthetic textiles. Platinum–rhodium alloys are used as catalysts in the production of nitric acid from ammonia. Thermocouples for measuring high temperature up to 1800°C consist of platinum with platinum–rhodium alloy. For still higher temperatures up to 2100°C the thermocouple wires are of iridium with iridium–rhodium alloy.

Rhodium is the easiest of the platinum group metals to electro-plate and has the advantages of white colour, high reflectivity and outstanding corrosion resistance. It is plated on high-duty electrical contacts, plugs and sockets and on white gold and silver jewellery.

RUTHENIUM. Like osmium and iridium, ruthenium is difficult to work and there is at present little use for it. With its relatively low density and price, however, it is an effective substitute for iridium as a hardening addition for platinum and palladium jewellery alloys. Ruthenium electro-deposition processes have been developed in recent years and, although the dark colour precludes the use of these deposits for decorative purposes, their hardness and corrosion-resistance approach that of rhodium.

SCANDIUM. The principal use of this rare element is in the manufacture of 'mixed metal halide' lamps based on scandium iodide. They give approximately 50 per cent more light than conventional mercury lamps of the same wattage.

SELENIUM. When light falls on selenium, its electrical conductivity increases by many orders of magnitude, depending on the intensity and wave length of the light. Selenium is not unique in this respect; all semi-conductors exhibit the effect to some extent. Advantage was taken of this property in the operation of early Xerox copy-machines. Temperature measurement devices, which base their action on registering the amount of light emitted by a hot material, rely on selenium.

SILICON. Although not a metal, silicon was bound to be mentioned many times in this book because of its important role in ferrous and non-ferrous metallurgy. It is a semi-conductor element but with chemical properties similar to those of carbon. As silicon occurs widely, in combination with other elements, particularly oxygen, in sands, clays, and other minerals, it is not surprising that when metals such as iron, aluminium, and copper are extracted from their ores, silicon often accompanies the metal as impurity. Pig iron made in the blast furnace contains as much as 3 per cent silicon. Most, but not all, is removed during the conversion to steel. In amounts of about one per cent, silicon confers improved elasticity to steels which also contain manganese and are used for car springs and bridges. The aluminium alloys containing silicon have been referred to on pages 157–60.

With the development of new chemical techniques and the procedure of zone refining, defined on page 298, silicon of exceptional purity can be produced, with the amount of impurities reduced to only one part in a hundred million. By doping this material with about one part in ten thousand of such elements as boron or phosphorous, transistors are made. The development of the silicon transistor made it possible to make drastic reductions in the size of computers and radios and also great improvements in their efficiency. Transistors used to be small objects that could be held in the palm of a hand but revolutionary decreases in size and cost were obtained by fabricating an entire set of integrated circuits on a silicon micro-chip, sometimes a hundred thousand transistors. As the efficiency of the silicon chip industry improved the products using them became more compact and cheaper. In 1970 a pocket calculator cost about £100; now, thanks to mass-production and technical developments, they are given away as sales gimmicks.

In some computers there are half a million transistors, all assembled on wafers of silicon; in very advanced computers the number of silicon integrated circuits may be several million. The silicon chip is included in the control of automobiles, trains, refineries, blast furnaces and space vehicles. Far too much pessimism has been generated about silicon chips causing unemployment; any young person with practical ability and an interest in electronics should note that there are widespread, exciting and well-paid opportunities for the many technicians who will make possible the wider use of silicon chips in information technology.

SILVER. Although this metal is resistant to many corrosive agents, it is attacked by sulphurous fumes which is the reason why silver articles tarnish and blacken in industrial atmospheres. Like gold, silver used for jewellery, teapots, cutlery and the wine labels one hangs on decanters is subject to hallmarking. In the United Kingdom there are two legal silver alloys, 'Britannia silver' containing 95·84 per cent silver and the much more common 'Sterling silver' containing 92·5 per cent silver. There is no restriction on the other alloying elements, but these usually are copper with or without cadmium.

English silver coins were first minted by Offa, King of Mercia in the eighth century, and silver remained in coinage circulation until 1947. However, for twenty-five years before that our silver coinage was being reduced in value as *Table 18* (page 222) shows, and the 'silver' coinage nowadays is a copper–nickel alloy.

A historic use of silver coinage remains. Maundy money, distributed by H.M. the Queen on the Thursday before Easter, contains 92·5 per cent silver, and 1, 2, 3 and 4p coins are minted for the occasion. The amount given is determined by the Monarch's age, one 'penny' for each year, and the number of recipients is governed by the same factor, one woman and one man for each year.

Industrial uses of silver include the lining of vats in dairy, brewing and other food processing, due to the metal's resistance to attack by organic corrodants. Silver has the highest reflectivity to light and the highest electrical conductivity of all metals, so it features in lighting reflectors, radar wave guides and for electrical contacts where high currents are involved. It had been expected that the industrial uses of

Table 18. Composition of British silver coinage

Years	Silver per cent	Copper per cent	Nickel per cent	Zinc per cent
1921	92·5	7·5	—	—
1922–6	50·0	50·0 (various silver alloys were used in this period)		
1927–47	50·0	40·0	5·0	5·0

Table 19. Free world silver consumption in millions of troy ounces
(*one million troy ounces equal about 30 000 kilograms*)

Use	1970	Anticipated 1980	Actual 1980
Industrial	280	520	340
Coinage	55	10	15

silver would nearly double between 1970 and 1980, as shown in the 'Anticipated 1980' column of *Table 19*; but the large temporary surge in the price of the metal at the end of 1979 caused the consumption to be much less than the estimate as shown in the final column. Even now the actual tonnage has not reached the estimated figure.

The free world uses about 13 000 tonnes of silver per annum, compared with a mine production of about 800 tonnes. The balance is provided by reclaimed silver from metal from the photographic industry. Most photographs depend on the light sensitivity of silver iodide, bromide and chloride. Much of the metal can be reclaimed and re-used.

SODIUM. This is a very soft, wax-like metal, which reacts vigorously with water and corrodes rapidly in air. Liquid sodium has proved valuable for extracting the heat which is evolved in the fast breeder reactor of which the prototype was built at Dounreay in the north of Scotland. Sodium melts at only 97°C and it is a superb material for conducting heat from the reactor core to the steam generator. The use of small amounts of sodium to modify aluminium–silicon alloys was described on page 157.

STRONTIUM. A reactive metal, resembling barium and calcium, strontium is also used in magnetic ferrites which were mentioned on page

204. It features in fireworks, burning with a crimson flame. The radioactive, lethal isotope strontium 90 is present in the fall-out from nuclear explosions.

TANTALUM. Though this element had been separated in an impure form early in the nineteenth century, it was not till 1903 that the German scientist Werner von Bolton isolated pure tantalum. From its initial limited use in metallic lamp-filaments, tantalum has grown into an important modern metal. Its three outstanding characteristics, high melting point, excellent corrosion-resistance, and ease of working, have assisted this growth. In its natural environment, tantalum is always found in association with niobium; a series of skilful chemical processes is needed to separate these two metals effectively. Tantalum metal, besides having the very high melting point of 2996°C, is unaffected by the majority of acids and consequently is used to avoid many corrosion problems in the chemical, pharmaceutical, and other industries. It is used in surgery for bone-splints, screws, brain clips, and other pieces which must be left permanently in the body. Because of its corrosion-resistance and its appreciably lower cost than platinum, tantalum is being employed in chemical equipment where heat has to be transferred under intensely corrosive conditions.

There has been a recent development of tantalum, in powder or foil form, for making small capacitors in the electronics industry. By using tantalum, it is possible to get capacitors of one tenth the size of the normal aluminium-foil ones, and still maintain the same capacity. The compound of tantalum and carbon is added to improve the performance of tungsten and titanium carbide cutting-tools. Tantalum carbide is the most refractory metallic substance, having a melting point over 4000°C.

TELLURIUM. When less than one tenth of one per cent tellurium is alloyed with lead it increases the hardness, strength, and corrosion-resistance. Tellurium is sometimes added to copper alloys and, in conjunction with lead, to steels to promote free-machining properties. Tellurium is not a true metal, but is more correctly described as a metalloid. It is mainly produced as a by-product.

THORIUM. This soft and ductile metal is closely related to uranium and is one of the very few naturally occurring elements which could possibly be utilized to develop atomic power by fission of one of its isotopes. Apart from future developments as nuclear fuel its main metallurgical use is for adding to magnesium to improve the metal's properties at high temperatures.

TUNGSTEN.* Robert Mushet, a friend and collaborator of Henry Bessemer, carried out experiments in making alloy steels and, in the years 1868–82, he discovered that steels containing tungsten were capable of cutting hard materials at high speeds while retaining a good cutting edge. At first Mushet made these 'self-hardening steels' with about 10 per cent tungsten, 1·2 per cent manganese and between 1·2 and 1·7 per cent carbon. Later he experimented with steels containing about 6 per cent tungsten, 1·5 per cent manganese, over 2 per cent carbon and about 0·5 per cent chromium. It was not till about 1900, thanks to the work of Fred Taylor and Maunsel White, that a really satisfactory steel for machining at high speeds was produced, containing 14 to 18 per cent tungsten. These have been described on page 150. A typical modern high-speed steel contains 18 per cent tungsten, 4·25 per cent chromium, 1·1 per cent vanadium and 0·75 per cent carbon. Traditionally many engineers in Britain prefer these tungsten steels, such as the one mentioned above, for cutting-tools; American engineers are using steels in which part of the tungsten is replaced by molybdenum. For example one such composition contains 6·4 per cent tungsten, 5 per cent molybdenum, 4·1 per cent chromium, 1·9 per cent vanadium and 0·82 per cent carbon.

Tungsten has the highest melting point of any metal (about 3400°C). It is extremely hard and it is about two and a half times as 'heavy' as iron. In modern times this combination of properties has led to the popularity of tungsten darts. The expert darts player requires a dart within the 26 to 28 gram range, but when these have brass bodies, they are stubby and cause difficulty in hitting a 'treble'. Elegant, slender dart bodies made by compressing tungsten powder with nickel, using processes described in Chapter 22, are only half as thick as the old-fashioned darts and are very suited to producing high scores, though they cost four times as much as the darts with brass bodies.

Tungsten carbide is one of the hardest known substances and has a wide range of uses, including tool tips for high-speed machining, the tyre studs of heavy vehicles that have to travel on icy roads, anti-personnel bombs and for the work rolls of the Sendzimir mill, described on page 83. Tungsten carbide is also used in making moulds for abrasive materials such as bricks, and for other moulds where great hardness must be combined with a very smooth finish, such as is necessary for making plastic sheeting.

---

* At a meeting of the International Union of Chemistry, held in Amsterdam in 1949, it was decided that the metal should be renamed 'wolfram'. Then a further meeting decided to revert to 'tungsten'. Now both names are officially recognized but the chemical symbol W is universally retained.

Wherever great hardness is required, whether for rock drilling or metal cutting, tungsten is likely to be used, either as a constituent of a high speed steel or a component of tungsten carbide, often blended with the carbides of other metals. Although tungsten is such a heavy metal, its widest use is connected with 'light', as the filaments of electric light bulbs.

URANIUM. The minerals pitchblende and carnotite contain uranium oxide U_3O_8. The U.S.A. has deposits in New Mexico, Utah, Wyoming, Colorado and Texas. Canadian ores come from the Elliot Lake area, and recent discoveries have opened the way for North Saskatchewan to become an important producer in the 1990s. A large mine in Namibia has been developed, while South Africa obtains uranium oxide as a by-product from gold mining. France augments her own resources from ores mined in her ex-colonial territories in Africa. Australia is becoming another major producer. In addition to mines in the U.S.S.R., ore deposits from East Germany enlarge the Soviets' stockpile. The Czecho-slovakian St Joachimstal mines, renamed Jachymov, provided Marie Curie with residue for her three-year task in isolating radium; now they supply the U.S.S.R. Calculating the richness of ores in terms of metal content, the amount of uranium ranges from 0·05 per cent in some American ores to 2·4 per cent in Australia's Northern Territory and about 4 per cent in North Saskatchewan. The metal, amounting to close on 60 000 tonnes per annum, is required principally for nuclear energy.

VANADIUM. The Swedish scientist Nils Gabriel Sefström was the first to deduce the existence of vanadium. During the 1830s he separated vanadium compounds from various minerals; some were black and others had attractive colours, so the metal, not yet isolated, was named after Fre Vanadis, the beautiful goddess of love in Scandinavian mythology. Forty years later a British scientist, Edward Thorpe, succeeded in making vanadium by treating its chloride with hydrogen. This metal was brittle and of little use and it was not until sixty years later that a purer and more ductile vanadium was made in America. Today the metal alone is hardly ever extracted but it is supplied mostly as the iron–vanadium alloy ferro-vanadium. This can be made in several ways; one typical process involves treating the compound iron vanadate with thermit – a mixture of aluminium powder and iron oxide which, when ignited, produces a very high temperature and reducing conditions which separate the vanadium from the compound.

The demand for vanadium has grown from a very small amount early in this century to over 30 000 tonnes per annum in 1987. The major sources of vanadium are open-cast mines in South Africa, the U.S.S.R., the U.S.A., Finland, Norway, Australia and China, where compounds of the metal occur alongside the iron ore magnetite. The supply has kept pace with demand and it is one of the few metals for which reserves appear to be ample for several more centuries.

The metal was included in steel to improve strength and hardness from the beginning of the twentieth century. An early high-speed steel contained 18 per cent tungsten, 4 per cent chromium and one per cent vanadium. The extra hardness which the metal confers comes from its carbide. The spread of vanadium steels in the automobile industry was partly due to Henry Ford, who observed that a French car that had crashed in an American race suffered a surprisingly small amount of damage. He helped himself to a piece of steel, had it analysed, discovered that the steel contained vanadium and decreed that future Ford cars should include vanadium steels. In the famous Model T Ford the crankshaft was of a chromium–vanadium steel.

The principal market for vanadium today is in the high strength low alloy (H.S.L.A.) steels. These contain only small amounts of vanadium – between 0·05 and 0·1 per cent, sometimes with even smaller amounts of niobium, but the strength of such steels is almost double that of straight carbon steels. Massive structures such as bridges, high rise buildings, spectator stands and aircraft hangars can be built economically, with the extra bonuses of reduced weight or longer spans because of the strength of the material. Pipeline operators can make cost savings by using H.S.L.A. steels because it is possible to increase the pressure and thus to convey a greater amount of gas. To take one of many examples, the 1300 km pipeline from Prudhoe Bay in the north of Alaska to Valdez in the south is 1·22 m in diameter and has a wall thickness of 16·6 mm. Six hundred and fifty tonnes of vanadium were required to produce a steel with 0·07 per cent of the metal for this pipeline.

YTTRIUM. This rare metal is now used in lasers in the form of an yttrium–aluminium garnet. It can give very fast pulses of the order of a few pico-seconds ('pico' indicating a millionth of a millionth). What is known as a magnetic bubble computer memory has been made from a similar garnet. Another recent application has been as a constituent of some nickel-based and cobalt-based superalloys; they have excellent resistance to corrosion and feature as coatings for turboblades in aircraft engines. Yttrium oxide, in combination with zirconium oxide, is used in the so-called Lambda-Sensors for determination of the oxygen content in automobile gases.

ZIRCONIUM. Although this metal was isolated over 150 years ago it remained a laboratory curiosity till recently. In the form of finely divided powder or swarf, zirconium ignites at low temperatures and finds application in photoflash bulbs, where it produces a very white light, making it suitable for colour photography. Because of its corrosion-resistance to alkalis, the metal is used in the rayon industry for spinnerettes. Its importance as an alloying element with magnesium is mentioned on pages 197–8.

Zirconium is valuable in nuclear engineering because of its low neutron absorption. The main outlet for zirconium is the alloy with 1·5 per cent tin, 0·1 per cent iron, 0·15 per cent chromium and 0·05 per cent nickel, used as a fuel canning material in water-cooled nuclear reactors, including those in submarines, and in structural components of nuclear reactors. Till recent years, zirconium as produced from its ores was contaminated with small amounts of the rare metal, hafnium. When the possibilities of zirconium in nuclear engineering were appreciated, it was realized that a hafnium-free metal must be produced, otherwise that impurity would cause slowing down of neutrons. Now good supplies of hafnium-free zirconium are available thanks to new methods of separating the metals by solvent extraction.

20

CORROSION

Corrosion is the destructive attack upon a metal by agents such as rain, polluted air, sea water or aggressive chemicals. The rusting of iron and steel provides the best-known examples of corrosion, and the continuous painting of steel bridges and ships illustrates that protection against rusting is an ever-present problem.

Many metal articles are used under conditions where they are affected by the atmosphere and by water; hence corrosion by the attack of these two media is the most familiar kind. Sometimes the conditions in which a metal gives service are of a severely corrosive nature. Oil platforms such as those in the North Sea have been described as factories on legs and, once in place, they have to remain in service for many years. Three fifths of an oil platform structure is permanently under the water, reaching down below the sea bed; another fifth is alternately covered and uncovered by tides and lashed by high seas with waves several metres high. Above and below the surface of the water are webs of pipes, nuts, bolts, rivets and welded joints, all subject to the vile conditions of the climate. The lower parts of the structure become the home of shellfish, seaweed and other forms of marine life. As soon as the platforms are installed on the sea bed the waves begin to drive against them, putting tremendous pressure on the materials from which they are made. As a platform is battered in this way, a small area of corrosion which has developed under a mass of seaweed may be the beginning of a crack and, just as bending a piece of wire backwards and forwards causes it to break, so the repeated impact of the sea may cause the crack to grow. In spite of these and many other problems, oil is reaching the mainland in substantial quantity. This success results from the skills of engineers and metallurgists who have developed the right grades of steels, the welding techniques, the platform designs and the metal protection systems.

Going from the sublime to the canine, the well known affection of dogs for lamp-posts causes city lighting engineers some harassing problems owing to the corrosive properties of the liquid with the formula K9P, which eats into the steel of lamp-posts a few inches above the ground. At one time there was talk of investing steel lamp-posts with a small electric charge that would provide a mild shock to persuade dogs to find a more suitable tree but the British sense of fair play triumphed and the dogs have not been penalized in this way.

The problem of corrosion may be approached by first considering the chemical action whereby the surface of a metal is attacked by gases in the atmosphere, which form chemical compounds with the metal. For example, one might imagine atmospheric oxygen gradually converting the surface of a piece of iron into iron oxide. However, atmospheric attack cannot be so simple as that, for if iron is stored for long periods in contact with dry air it does not rust so much as would be expected from the usual behaviour of the metal. If the iron is kept moist, rusting proceeds rapidly, which might lead one to expect that water was the influence causing the corrosion. But in air-free distilled water rusting does not take place so fast as when air is dissolved in the water. Even the combined presence of air and moisture does not fully account for the formation of rust, for pure iron is attacked less than commercial grades of iron or steel. These observations led to the realization that corrosion is stimulated when a metal is non-homogeneous and when it is in contact with water containing gaseous, liquid or solid substances.

Various theories have been put forward to account for the vagaries of corrosion, but facts such as those given above confirm that a relationship exists between corrosion and electrolysis. In other words, a corroding metal behaves as though it is an electric cell or wet battery. At first the relationship between the rusting of a nail and the action of a battery may not seem obvious, but it has been proved that corrosion is accompanied by the setting-up of small localized electric currents. In a battery the current is produced by suspending two different metals in a chemical solution. When the circuit is completed, one metal, identified as the anode, dissolves, while an electric current flows through the solution from this metal to the other, which behaves as the cathode. One may imagine a piece of metal corroding in this way; the presence of particles of an impurity, or contact with some other metal, allows a difference of voltage to be set up, thus causing a minute electric current to flow. The moisture that is present, containing air or some dissolved chemical substance, conducts the electricity and corrosive attack is begun.

When two metals are in contact, in conditions where electrolytic action may take place, corrosion of one metal will occur. This effect, known as galvanic corrosion, was recorded over two hundred years ago. The Royal Navy frigate *Alarm* had her hull covered with copper to prevent damage by barnacles. After only two years' service it was found that the copper sheathing was detached from the hull in many places because iron nails that had fastened the copper to the timber had become severely corroded. Some nails, which were much less corroded, had been insulated from the copper by brown paper which was trapped under the nails' heads. The copper had been delivered to the dockyard wrapped in this paper, which had not been removed completely when the sheets were nailed to the hull. It was, rightly, concluded that iron should not be allowed to be in direct contact with copper in sea water. This type of galvanic corrosion was allowed to occur many times because designers and engineers did not realize the havoc it would cause. As recently as 1982, during the Falklands War, two Sea Harriers suffered nose wheel collapse because of galvanic corrosion between the magnesium alloy wheel hubs and their stainless steel bearings.

Metals and alloys can be listed in what is known as a galvanic series which helps to predict the likelihood of corrosion of pairs of dissimilar metals in given corrosive media. Magnesium and zinc are at the top of the list, showing that they are likely to be sacrificially corroded when in direct contact with other metals or alloys. Titanium, nickel–chromium alloys and platinum are at the bottom of the list, showing that they will be protected by the more active ones. North Sea oil platforms are massively protected by zinc sacrificial anodes welded over the submerged parts of the platform. In several cases as much as 30 per cent of the submerged mass is zinc anode material.

The great difficulty in studying corrosion is that so many variable factors influence its initiation, course and final result. For example the purity of the metal, the composition and inter-relation of all the substances with which it comes into contact, the presence of bacteria and the possible presence of small stray electric currents may all affect the corrosion.

Sometimes the corrosion behaviour inside a crack or recess in a metal article provides an example of what is known as 'differential aeration'. Moisture which has penetrated into the bottom of the crack becomes devoid of dissolved atmospheric oxygen, while at the mouth of the crack the water contains plenty of dissolved oxygen. This provides an electrical potential difference, that is an anode at one end and a cathode at the other. Under such circumstances, the lower areas devoid of oxygen become the anode and they corrode.

METHODS OF COMBATING CORROSION

The more that can be learnt about the causes and mechanism of corrosion the easier it will be to prevent its damaging action; there are several possible ways of doing this. Pure metals are likely to resist corrosion better than metals containing impurities; thus pure aluminium resists attack better than a commercial variety; 'chemical lead' of high purity is more resistant to corrosion than a less pure grade. The elimination of impurities from alloys usually helps to improve their corrosion-resistance; also, in a given series of alloys, those which contain only one structural constituent are likely to resist attack better than those which contain more than one constituent.

The search for alloys with superior resistance to corrosion is a never-ending attempt to satisfy an insatiable demand. In ordinary structural steels quite small additions of copper of the order of 0·05 per cent improve the resistance to corrosion. Such improvements, though significant, are nothing like so powerful as the effect of additions of 18 per cent chromium and 8 per cent nickel, which makes steel stainless, that is to say, rustless, under most normal atmospheric and other corroding conditions.

If the initial coating of corrosion product is adherent and impervious it can prevent further attack from taking place. Aluminium provides an illustration of this, for it forms a thin but strong film of aluminium oxide on its surface which restricts further corrosion. Stainless steel containing chromium behaves in a similar manner, its corrosion-resistance being due to the formation of a protective oxide film. On the other hand, ordinary iron and steel are liable to progressive rusting because the layer of rust on the surface is porous and tends to flake off, so further rusting penetrates inwards, though often at a reduced rate.

The metal may be coated, to prevent the access of corrosive agents. For example, protective coatings of other metals can be applied by electro-plating, dipping, spraying, or by cladding one metal with another, as when sheets of strong aluminium alloy are clad with pure aluminium (described on page 165). Most iron and steel, however, is protected by paint, bitumen or proprietary preparations. During the past generation the amount of maintenance on bridges has been much reduced, thanks to a better understanding of the mechanism of corrosion, carefully organized codes of practice for treating steel structures, and the development of improved methods for protecting steel including spraying coatings of zinc or aluminium. Nevertheless, it is difficult to protect the underneath surfaces of river bridges; the moisture that is present on the steel for most of the year induces the breakdown of coatings and makes it hard to obtain the almost clinical conditions required for optimum performance of the protective treatment.

The use of synthetic resins in paints has increased enormously, thus giving a long life to the paint film; finishes consisting of plastic materials and rubberized paints have also been introduced. Each type of finish or protective coating has its own particular aesthetic or economic merit. One of the most recent developments in corrosion protection has been the use of inhibitors, which are chemicals put into cooling and central heating systems and in anti-freeze solutions in car engines. For this purpose, chemicals such as phosphates, silicates, chromates and benzoates are widely used.

CATHODIC PROTECTION

It is sometimes possible to reduce corrosion by a method called 'cathodic protection', illustrated by the galvanizing or zinc-coating of mild steel. The zinc coating slowly disappears but it has delayed the rusting of the steel. This effect is also described as 'sacrificial protection'. A similar system operates for the protection of underground pipe lines. These are usually wrapped with impregnated cloth or given a bitumen coating, but it is often necessary to provide additional protection at gaps in the wrapping or at pores in the bitumen. This may be achieved by burying zinc or magnesium anodes in the soil close to the steel pipe and connecting them to it by an insulated wire. Current flows along the wire to the other metal which corrodes instead of the steel and thus has to be replaced periodically.

A third form of cathodic protection is applied, particularly to steel structures in sea water; a direct electric current, closely controlled in potential and amperage, is passed through the sea water to the steel. The protective current has to be conveyed through an insoluble anode, usually of graphite, lead, iron–silicon alloy, or through a titanium rod or plate thinly coated with platinum. Many of the piers and jetties erected for mooring and unloading oil tankers are protected in this way; for example at Thames Haven.

In contrast with such methods of cathodic protection, it is possible to build up an oxide surface layer by making the metal the anode in a special electrolyte, when a controlled current is passed. The oxide layer so formed is tough, and resistant to wear and corrosion because the underlying metal is insulated both electrically and physically. Apart from aluminium, other metals can be anodically treated, including titanium and stainless steel.

CONDENSERITIS

'Our ships, great and small, have been at sea more continually than was ever done or dreamed of in any previous war since the introduction of steam. Their steaming capacity and the trustworthiness of their machinery are marvellous to me, because the last time I was here one always expected a regular stream of lame ducks from the fleets to the dockyards, with what is called "condenseritis", . . . but now they seem to steam on almost for ever.' These words were spoken by Winston Churchill when he presented the Navy Estimates to the House of Commons on 27 February 1940, but probably few appreciated that his remarks paid tribute to years of painstaking research work by a handful of British firms. In all forms of steamships, a condenser is vital to secure maximum efficiency; it consists of nests of tubes about 4 to 7 metres long and 18 mm in diameter, with walls about 1·5 mm thick, all arranged in a gigantic honeycomb. Cold sea water is circulated through the condenser tubes while, outside the tubes, exhaust steam from the engines condenses to pure water and is returned to the boiler for generating more steam. If the tubes corrode, the sea water leaks into the boiler water, causing rapid corrosion of the whole power unit.

During the First World War the average life of condenser tubes, which were then usually made of 70/30 brass, was only a few months and in the early years of the Second World War many naval vessels were immobilized because of failures in the condenser tubes. In 1940 the developments mentioned by Winston Churchill were taking place. At first a copper alloy with 5 per cent nickel and one per cent iron gave a much improved condenser life and it had the advantage that the existing copper-smithing techniques could be used. During the following years the flow velocity of the cooling water was nearly doubled and the 5 per cent alloy could not resist the eroding effects of the water's turbulence. A copper alloy with 10 per cent nickel was introduced for surface vessels and one with 30 per cent nickel for submarines.

The number of steamers declined in favour of oil-fired ships that required condensers that would withstand severe corrosion. Then desalination plants and the equipment of the offshore oil industry demanded the solutions to further problems. Initially oil platforms in the North Sea had galvanized steel for piping systems conveying sea water but rapid failures due to corrosion led to expensive replacements and repairs. The copper alloy with 10 per cent nickel was introduced and gave a greatly improved life. The metal cost was about a third higher than that of the galvanized steel but that was insignificant compared to the savings in downtime and repairs. In conditions where the cost of breakdown would justify an even better resistance to corrosion, titanium is used and one company has supplied tens of millions of metres of thin-wall titanium tube for steam-turbine condensers.

STRESS-CORROSION

Failures due to stress-corrosion cracking or corrosion fatigue take place at stresses well below the normal tensile limit of the metal involved. As the names imply, in the first case fracture results from the combined influence of corrosion with tensile stress, while in the second case rapidly reversing stresses under corrosive conditions lead to failure.

For a given metal system there is usually a specific corrodent or series of corrodents capable of initiating cracking. Thus stress corrosion in brass takes place in the presence of ammonia, and for stainless steel the presence of chlorides constitutes the hazard. For mild steel, contact with strong caustic soda can result in 'caustic cracking' which sometimes affects the rivets of high-pressure steam boilers. The caustic agent comes from alkaline boiler waters, which can seep into and concentrate in the crevice under the rivet head. There seems little doubt that the initiation of cracking results from a process of corrosion on a microscopic scale.

Stress-corrosion failures fall into two categories: those taking place in the presence of stresses introduced into the metal during a deformation process and remaining 'locked up', and those stresses imposed by external forces applied during assembly of equipment or during its operation. Locked-up stresses can be removed by a low-temperature annealing process which results in more even distribution of the stress within the metal without lessening its strength.

Examples of stress-corrosion from locked-up stresses occurred in India in the late nineteenth century. British infantry, artillery and cavalry were present in force and were supplied with ample ammunition. During the wettest months of the monsoon season ammunition was stored in stables till the dry weather returned but many brass cartridges were found to have suffered what became known as 'season cracking' which led to bad accidents when cartridges 'blew back' when fired. The cracking seemed to be due to some form of corrosion, though the actual attack on the metal was only slight. Other examples of season cracking were noticed and it was remarked that they occurred only when a metal had been cold-worked in its fabrication and later subjected to corrosive conditions. It was not till 1921 that the cause of season cracking was discovered. The cartridges had been subjected to stresses in their manufacture, as will be remembered from the description of the making of a brass cartridge case, on page 173. Corrosive conditions were provided by the combination of humidity in the atmosphere plus horse urine in the stables. When the mystery was solved it became possible to remove the pent-up stresses in the brass by annealing at a temperature of 200°C to 300°C. By this treatment the stress is relieved, so that the brass will not tend to season crack, and the annealing temperature is not sufficiently high to reduce the mechanical strength of the metal.

EROSION AND CAVITATION

Loss of metal by surface damage is often included in studies of corrosion. Some parts of marine propellers, water turbines, hydraulic and hydro-electric pipe lines suffer from cavitation attack caused by the breakdown of streamline flow of the liquid passing over the surface of the metal. Such sudden flow changes produce 'vacuum bubbles'; where these bubbles collapse, water-hammer blows occur which progressively damage the metallic surfaces. This usually manifests itself as pockmarks, where metal has been eaten away.

Condenser tubes in power stations are often thinned internally by high velocity steam or water. Such effects may be increased in severity by particles of scale or sand which have become trapped in the tube. The formation of eddy currents and turbulence can also lead to the erosion of liquid-carrying tubes. High velocity wind carrying dust or particles of salt can cause erosion in aluminium overhead conductors or other exposed structures.

MICROBIOLOGICAL CORROSION

A micro-organism can be viewed as a catalytic entity which may produce high concentrations of such compounds as hydrogen sulphide or organic acids from materials containing sulphate ions or petroleum hydrocarbons. The most important group of bacteria associated with the corrosion of metals are the sulphate-reducing bacteria. They obtain their carbon and energy from organic nutrients and their respiration utilizes sulphate ions which results in the formation of highly corrosive sulphide products.

CORROSION TESTS

The behaviour of metals and alloys in service can be forecast by subjecting them to accelerated corrosion tests. As a result of years of experience, these are fairly accurate, provided that control factors are taken into account. Such corrosion tests are usually conducted in closed chambers with water sprays containing sodium chloride, hydrogen peroxide, or acids, whichever causes the appropriate type of corrosion attack. In other tests sprays may alternately wet the specimen and dry it with warm air.

In some corrosion tests, liquids are made to impinge rapidly on portions of tube or flat specimens which are strung together, and electrically insulated from each other. The specimens are weighed before and after test; they are examined for the type of attack, whether general thinning or localized. In other tests, jets of liquids impinge on the samples of various metals and alloys being examined. As a result of many such tests, it has been found that the flow of a liquid can be quite critical, according to whether it is streamlined or turbulent.

Corrosion tests, therefore, help to predict when corrosion is likely to occur, but they should be regarded as only one of the weapons needed in the everyday fight against corrosion. Nowadays the chief concern of corrosionists is to 'get the message across' to the architects and designers on the job, so that they will take advantage of all the knowledge that has been obtained on the prevention of corrosion. It is rather striking that the growth of knowledge about this complex subject has caused an important change of attitude. At first corrosion was regarded as an unavoidable evil and the best that could be done was to build structures with more than ample weight of metal and apply paint liberally and regularly. Now metallurgists, corrosion specialists, designers and engineers have gone into the attack, with the object of preventing corrosion from taking place. The fight against corrosion starts on the drawing board.

21

JOINING METALS

When a metal component is being intelligently designed many factors are taken into account: the cost of the finished article, its strength, reliability in service and appearance. Often the facilities and skills of the manufacturer cause one or other process to be selected. Sometimes a component is designed in one piece and processes such as diecasting or investment casting are used. In contrast to these methods it is often found that components can be made effectively by assembling cheaply produced shapes of simple design, pressings for example. The attention which has been paid in recent years to joining processes, such as soldering, welding and joining by adhesives, has made 'built-up' construction both rapid and efficient, and it is often cheaper than 'one-piece' manufacture. There has been spectacular progress in the joining and assembling of metal articles, from the making of biscuit tins to the building of ships. In this chapter the main types of joining processes are grouped into mechanical methods, soldering, brazing and finally welding.

MECHANICAL METHODS

The Bayeux Tapestry shows a fine display of armour and illustrates some ancient methods of joining metals. The sheet iron helmets were joined by riveting to the 'nasals' which protected the warriors' noses. The armour worn at the Battle of Hastings was fashioned by solid welded rings alternating with riveted rings.

During and after the Industrial Revolution, the process of riveting was widely and noisily used, from Brunel's *Great Eastern* to the *Queen Mary* nearly a century afterwards. A number of well-known structures provide other examples of the use of riveting, from the Statue of Liberty with a mere 300 000 to the Eiffel Tower with $2\frac{1}{2}$ million and the Forth railway bridge with 7 million rivets.

High bond-strength adhesives have become available for joining metals. They now replace bolting, riveting, soldering and welding, in many fields of engineering. These adhesives are strong, economical and versatile, and they enable metals to be bonded with many materials, including other metals, glass, ceramics, wood, concrete and plastics. Because the whole bonded area is covered with adhesive, there is no bi-metallic corrosion. In the automotive industry aluminium alloy body structures are assembled by using heat-treatable alloys and adhesive bonding. The paint stoving cures and strengthens the adhesive bond and age-hardens the alloy. This technique is being used by Rover, Volvo and Fiat.

SOLDERING AND BRAZING

The operation of soldering consists of uniting metal components by a metal or alloy which melts at a lower temperature than either piece of metal to be joined. When liquefied the solder covers the heated surfaces and forms an alloy layer less than a tenth of a millimetre thick, so that when the parts have cooled the two pieces of metal remain firmly united. Soldering processes for joining gold and silver work can be traced as far back as 4000 B.C.; the 'solders' were alloys of gold with copper, or silver with copper. Nowadays the most familiar solders are those containing mainly lead and tin, which have been discussed on pages 42 and 179. These can be melted with a blow lamp or heated soldering iron (the 'iron' actually being of copper). Metals which are to be soldered should be cleaned before commencing work; and, in order to maintain the clean surface and to remove oxide films, a flux of resin or zinc chloride is usually applied which further cleans the metal surface and seals it from the tarnishing effect of the atmosphere.

The ease with which a solder will wet and flow over the surface of another metal depends on the characteristics of the metal concerned. For example, molten lead will not wet a copper surface, however clean it is, but if a little tin is added to the lead the resultant alloy will readily flow over copper. Aluminium used to be regarded as a difficult metal to solder because the tenacious film of aluminium oxide which forms on the surface interferes with wetting. Similar problems arise with many other alloys containing aluminium; as little as 0·5 per cent aluminium makes soldering somewhat difficult. Special fluxes have been developed for soldering aluminium and, although some care is necessary, satisfactory joints can be made. Considerable use has been made of this development in the electrical industry, for making cable joints in aluminium wire cables.

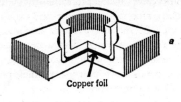

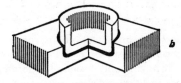

Copper foil

Fig. 46. Capillary jointing

For many years bicycle frames have been made by joining steel tubes to the brackets by brazing operations, the techniques becoming more sophisticated and faster over the years. Dip brazing, in which parts are immersed into molten brass, has disappeared from the modern cycle works; most assemblies are now joined by preloading internal charges of brazing solder into the assembly and applying external heat.

The brazing solder is a form of brass (hence the word 'brazing') which usually consists of 60 per cent zinc with 40 per cent copper. The alloy can be melted at about 850°C by a gas torch. Such higher-melting-point solders are often called 'hard' solders, in contradistinction to the lead–tin or 'soft' solders, which melt at 200°C to 250°C. A brazed joint is considerably stronger than a soft-soldered joint and, partly for this reason, there has been a tendency towards brazing instead of soldering, a trend which received further impetus owing to the high cost of tin.

One technique which has arisen from the development of brazing methods has been capillary jointing, a method used in the manufacture of motor-car accessories, refrigerator parts, and radio sets. If two steel parts are to be joined as shown in *Fig. 46*, they are placed together with a thin disc of copper foil between them (*a*). The parts are then passed through an electrically heated furnace filled with an atmosphere free from oxygen but containing hydrogen. The copper foil melts, but is kept clean and free from oxide film by the presence of the hydrogen and so the molten copper flows in the gap; capillary attraction, better known as surface tension, pulls the molten metal up the sides (*b*). This process lends itself to mass-production methods; the join is clean, neat and economically made.

Silver solders are alloys of copper, zinc, and silver, sometimes with small additions of cadmium; one solder contains 60 per cent silver with 30 per cent copper and 10 per cent zinc, the melting point of this alloy being 735°C. At the beginning of this century silver solders were used mainly for jewellery but in the 1930s several silver alloys were developed for engineering use on account of their strength, fairly high melting point, and the accuracy and neatness of the joint obtainable.

WELDING

Welding processes can be classified into two groups, namely those involving pressure (generally accompanied by melting or heating nearly to the melting point) and those involving only fusion. The former group includes both forge- (or hammer-) welding and various resistance-welding methods, while the latter is concerned with methods utilizing heat sources comprising oxy-fuel gases and electric arcs. Some metals will weld even without heat being applied; for instance, two clean flat pieces of lead, silver, gold, or platinum can be laid face to face and hammered cold until they become welded. Each metal or alloy has a temperature above which hammer-welding becomes possible, though oxidation of the surface of the metal often interferes with the success of welding at this minimum temperature. For example, hammer-welding of aluminium is difficult, because the layer of oxide prevents a metal-to-metal contact; cast iron is practically impossible to hammer-weld because it is brittle.

Chains used to be made by strenuous manual effort; each piece of hot iron bar was threaded through the preceding link, turned over at the ends and made into a link by hammer-welding.* Nowadays, chains with links up to 40 mm diameter are made automatically, the round bars being formed and bent to shape and then welded in a second automatic operation. Chain links of over that size are shaped on hydraulic bending machines and then welded.

The medieval sword-maker had to be an expert in hammer-welding. He was faced with the problem of combining the sharp cutting edge obtained in a high carbon steel with the need to prevent the sword from shattering if it was too brittle; but if the sword was too soft it would bend and be equally useless. The solution, discovered in many parts of the world in ancient times, was to hammer-weld layers of metal together, a soft iron or low carbon steel in the centre and a much harder high carbon

* The Black Country Museum, at Dudley in the Midlands, has numerous fascinating exhibits of 'old-time' industries, including a demonstration of chain-making. Well worth a visit!

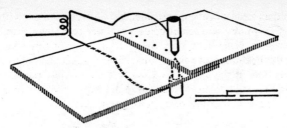

Fig. 47. Spot-welding

steel to provide the cutting edge. All the pieces would be hammer-welded together to make a sword blade with the required properties; then the weapon would be fashioned and decorated most exquisitely. In the third century a process now called pattern-welding was used in which iron rods and wires were twisted together, heated and forged so as to join them by welding, and some very decorative swords were made in this way.

ELECTRIC RESISTANCE WELDING

When an electric current is passed through two pieces of metal in close contact they heat up, depending on their electrical resistance; heat also develops at the points of contact. This is applied in a process known as resistance butt-welding. The two metal parts are placed in separate current-carrying electrodes and butted together while a high electric current at a low voltage passes across the joint. Heat is developed at the point of contact and when the welding temperature is reached, pressure is increased and a joint is formed between the two components.

Another method, known as spot-welding, illustrated in *Fig. 47*, is particularly suitable for joining sheets of metal which are held together with two electrodes lightly pressing on either side. An electric current is passed through the electrodes, thus heating the metal between them. Immediately afterwards the two electrodes are pressed tightly together and the metal sheets are joined effectively. Spot-welding is used for joining steel-sheet fabrications such as washing machines and refrigerator cabinets. Motor vehicle bodies contain several thousand spot-welds and some aircraft frames contain tens of thousands.

A further development of spot-welding is seam-welding, in which a pair of metal sheets is passed between electrodes in the form of rollers, so that a continuous line of weld is obtained. *Fig. 48* illustrates

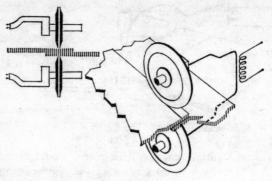

Fig. 48. Seam-welding

seam-welding, though in some equipment the lower electrode may be replaced with a flat bar.

Electrodes must have good electrical conductivity and are usually of a copper alloy such as copper–chromium, copper–chromium–zirconium and occasionally of copper–cobalt–beryllium or copper–nickel–phosphorus. The electrodes need maintenance after about ten thousand welds and then they may be re-faced.

In ultrasonic welding two metals, not necessarily of similar composition, are placed in close contact and subjected to high frequency vibrations. Special equipment is required to convert electrical energy into rapid mechanical vibrations and the process is used for bonding lead-out wires for semi-conductor devices and encapsulation of explosives.

OXY-FUEL WELDING

For welding joints that are larger than those achieved by spot-welding, additional filler metal must be introduced between the two pieces to be joined. The filler metal and part of the parent materials are melted and allowed to fuse together, thus uniting the two metal components. The essential difference between this process and brazing is that in brazing only the filler metal is melted. Oxy-fuel gas welding can be done with propane or butane, but acetylene has advantages over those gases because it provides a higher flame temperature, about 3000°C, and it uses less oxygen (only one third that required for propane).

By adjustment of the supply of oxygen or acetylene from the gas cylinders, the type of flame can be regulated according to the require-ments of the metal to be welded. Some metals, for example brass, weld better if the amount of oxygen is in excess of that required to burn acetylene, while cast iron welds better in a reducing flame in which surplus acetylene is present.

The flame is played on the parts of the metal to be joined and also on a metal filler rod which melts and runs into the gap between the two metal parts and joins them. A flux has to be used in welding some metals, in particular aluminium, though it is not necessary for steel.

ARC WELDING

The metallic arc process is used for the welding of ships' structures, large machines, structural steelwork and pipelines (*Plate 19a*). The heat is developed by striking an electric arc between the work and a metallic filler rod. This is usually of a low carbon steel, but additions of metallic powders and deoxidizers to the flux coating modify the composition of the weld metal. The formation of the arc develops so much heat that the end of the rod melts and forms a weld. The rod is coated with suitable flux which melts when the arc is struck, so that the weld solidifies and cools under the protective coating of molten flux.

There are several variations of arc welding. In the tungsten inert gas (T.I.G.) process, sometimes called argon arc welding, an arc is struck between the component to be welded and a tungsten electrode held in a suitable container, around which flows a stream of inert gas, usually argon. This protects the heated tungsten and the welded metal from oxidation, thus dispensing with the use of fluxes. T.I.G. has revolu-tionized the joining of non-ferrous metals and has made it easy to join aluminium, which used to be difficult to weld because of the oxide skin which so readily forms on the metal. It has also become possible to weld titanium, zirconium and tantalum, thus helping to widen the use of those metals in aircraft, chemical engineering and other industries. The T.I.G. process is widely used for welding stainless steel for nuclear reactors and chemical plant.

When welding any aluminium alloy an alternating current arc must be used. The arc exerts a cleaning action on the surface of the aluminium which removes the thin tenacious film of aluminium oxide and enables the edges of the materials to melt and flow together. With T.I.G. welding there is virtually no limit to the thickness of material that can be joined but the process is quite slow and labour-intensive and may therefore be of limited economic appeal. T.I.G. welding can be used in all positions, including vertical and overhead situations. It is popular for the preliminary welding, known as the 'root run', of in situ pipes, where the workpiece obviously cannot be rotated. After T.I.G. deposition of the root run, a cheaper process such as manual metal arc is often used to complete the weld. Indeed, some T.I.G. machines can be changed to metal arc at the touch of a switch.

In a variation of T.I.G. the metal inert gas process (known as M.I.G.) maintains an arc between the end of a base wire electrode, usually copper-coated mild steel, and the components to be welded. The wire is fed automatically at constant speed and the weld is shielded by an inert gas, such as argon, or an 'active' gas, such as carbon dioxide, or a mixture of the two.

The submerged arc process is specially used for high quality pressure vessels, chemical plant, structural engineering and shipbuilding. The electric arc is maintained between the end of a bare wire electrode and the work. As the electrode is melted it is fed in the arc by a set of rolls, driven by a motor. The arc operates under a layer of flux (hence 'submerged' arc).

The various techniques of arc welding offer rapid and effective methods of construction in shipbuilding. By the use of welding instead of riveting weight is saved, since riveted joints require overlapping of the ship's plates and each rivet adds its little load to the dead-weight of the vessel. Large and small naval vessels have been made with extensive use of arc welding. Ten million rivets were required in the *Queen Mary*, built in the 1930s, but thirty years later the *Queen Elizabeth 2* was a welded structure. The only rivets were those used to connect the aluminium superstructure to the hull.

NEW DEVELOPMENTS

In electron beam welding the work piece is heated by the bombardment of a beam of electrons. Filler metal may be added, if necessary. The work is usually done in a vacuum chamber, but electron guns are being developed which allow the electrons to travel a short distance through air. This process is used for dealing with difficult-to-weld materials, and for such specialized products as turbine-blade root joints.

Diffusion bonding is used for welding heavy section constructions, hollow assemblies and other inaccessible joints. The parts to be joined are machined, cleaned and clamped together, then placed in a vacuum chamber and heated. Bonding normally occurs at about 70 per cent of the melting temperature of the material and after a few minutes a diffusion bonded joint is produced, with good mechanical properties.

Magnetically impelled arc butt-welding, known as M.I.A.B., is used in the mass production of tubular assemblies such as automotive back axles. Two steel tubes are clamped so that there is a gap of about 2 mm between them. One is advanced and retracted to form an arc, which is rotated round the joint interfaces by the interaction of the magnetic fields set up by the coils and the arc current. At the conclusion of the timed cycle, about 6 seconds, the ends have become forged together.

A development which has grown from an exotic laboratory technique in the early 1970s to a present-day practical manufacturing tool is known as plasma-arc, or plasma-jet welding. In this type of torch high-energy inputs are possible, up to 50 kW, and alloys can be fed in as powder; materials such as borides, carbides and nitrides, which have very high melting points, can be melted and sprayed on to a backing material such as steel. One example of the scope of this process is the 30 km of plasma-welding that was carried out to line concrete tanks with stainless steel sheet for a copper-extraction plant in Zambia. More recently the plasma-arc process has been extended by what is known as Plasma-M.I.G. welding, a combination of plasma-arc and argon-arc welding.

The friction-welding process that was first developed in the U.S.S.R. about thirty years ago is now being used by all industrialized countries. One metal is rotated at high speed and then pushed hard against another, causing the two to be welded. Such friction welds can be made between dissimilar metals; a typical example is in aluminium refineries where they use aluminium conductors. These cannot withstand the very high temperatures close to the furnace so at that point steel is used, friction-welded to the aluminium. A recent development, radial friction welding, involves an annular ring being rotated around a butt joint; this process is particularly applicable to the joining of shafts to collars or joining pipes on site.

In laser welding the laser generates a powerful parallel beam of light, which is focused to a point at which the entire output of the laser is concentrated. The Ford Motor Company in the U.S.A. has been using laser welding since 1975.

Fully automated welding by robots is already with us. Fiat in Italy were pioneers and a production line in Turin features the assembly of car bodies by means of automatic spot-welding. Another impressive achievement is in the shipyards of Japan at the Ariake yard of Hitachi, where more than three quarters of the fabrication of supertankers is by means of automatic welding. *Plate 19b* shows automatic welding by a robot.

UNDERSEA WELDING

The difficult and dangerous task of undersea welding of oil rig platforms, bracings and pipelines offers serious challenges to the ingenuity of welders and welding engineers. There are three basic techniques which have been adopted. The simplest and cheapest is 'wet' welding, which requires no specialized equipment and generally involves metal-arc, metal inert gas or flux-cored arc welding. However, high quench rates due to the surrounding water coupled with the pick-up of hydrogen into the weld may lead to poor quality welds of inadequate strength and toughness.

Hyperbaric welding,* illustrated in *Fig. 49*, involves building a chamber round the joint to be welded, then displacing water from the chamber by gas pressure. The pressure inside the chamber is therefore equal to that prevailing outside at the depth of the chamber. Hyperbaric welding has become an accepted method of subsea repair and maintenance welding, down to water depths of about 300 metres. At greater depths than this, problems are experienced because of the high pressure inside the chamber; replacement of human welders with automated or robotic equipment is one possible solution.

Atmospheric welding can be performed at depths below which saturation diving is not possible. In this system a working pressure equivalent to the atmosphere is maintained. The system is based on the use of a disposable pressure vessel or work chamber, linked to a second, re-useable, diving sphere that contains all the equipment necessary for welding and its related processes. This is the most costly of the three methods.

The latest developments within the welding industry have aimed at increasing productivity and quality by improving the power source response, accuracy and control. Such improvements have led to the increasing adoption of mechanized, automated and robotic welding in many industries. A completed weld, whether deposited by man or machine, is dependent on the skill of the welder, or the programming of the machine and expertise of the welding engineer. A weld has been described as the whole of metallurgy in miniature, for it calls for a knowledge of the melting, casting and forging of metals. A weld carried out by an inexperienced operator, or by an incorrectly programmed robot, may be weak, brittle, coarse-grained and unsound. On the other hand skilled operating and planning can produce welds that are at least as strong as the parent metal. The weld can be proved to be satisfactory and reliable by inspection techniques such as radiography and ultrasonic testing.

* This and other technical terms are defined in the Glossary, pages 292–8.

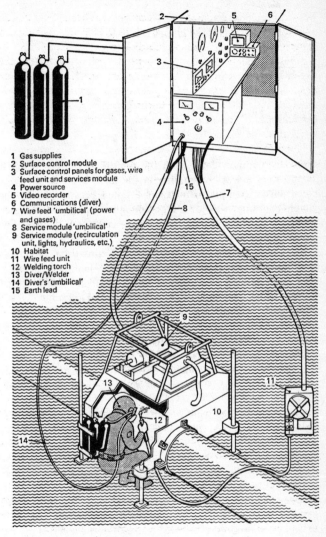

1 Gas supplies
2 Surface control module
3 Surface control panels for gases, wire feed unit and services module
4 Power source
5 Video recorder
6 Communications (diver)
7 Wire feed 'umbilical' (power and gases)
8 Service module 'umbilical'
9 Service module (recirculation unit, lights, hydraulics, etc.)
10 Habitat
11 Wire feed unit
12 Welding torch
13 Diver/Welder
14 Diver's 'umbilical'
15 Earth lead

Fig. 49. Undersea welding
(*Courtesy B O C SubOcean Services – a B O C company*)

22

POWDER METALLURGY

Interest in platinum dates from the middle of the eighteenth century, although then it was known only in the form of its naturally occurring small grains. Attempts to melt these failed because the furnaces then available could not provide a high enough temperature (the metal melts at 1735°C). At the end of the eighteenth century William H. Wollaston discovered that larger, malleable pieces of platinum could be made by pressing the powdered metal into blocks, heating them in a coke furnace, and striking them while hot with a heavy hammer. Later a similar process was applied to tungsten, which could be produced in powder form, but which even today cannot be melted on a commercial scale.

From such beginnings powder metallurgy developed; it has provided new ways of mass producing finished, accurate metal components which could not be made economically by any other means. Although a considerable quantity of non-ferrous products such as bronze bearings and magnetic alloys are made by powder metallurgy, by far the largest amount – about 300 000 tonnes per annum – is produced from ferrous metals.

It may be wondered what advantages are to be derived from changing metals into powders if they are intended to be compressed together again. There are several:

1 Some metals cannot be melted under commercial conditions, though they can be made as powders by chemical processes and then formed into solid pieces, such as the tungsten filaments of electric light bulbs.

2 Powder metallurgy offers an economical way of shaping components, first by compacting the powder and then by sintering. This is a building-up process where practically all of the metal is used and there is no machining scrap. In these respects the process is at least 95 per cent efficient.

3 Non-metallic substances and intermetallic compounds, which cannot

be incorporated into metals by casting, can be added into a powder mix. Graphite, an excellent lubricant, can be put into bronze which is to be used for bearings and carbon incorporated into copper for current-carrying bushes in electric motors. The biggest single outlet in this category is the production of tungsten carbide tool tips.

4 Metal powders can be mixed to form alloys: powders of copper and tin in the correct proportions are compacted and sintered at about 800°C, producing bronze. A new process called mechanical alloying involves ball-milling two materials, such as yttrium oxide in a nickel-base alloy, to form a composite material.

PRODUCTION PROCESSES

Metals can be produced in powder form by three principal methods. In atomization molten metal is broken up into droplets and rapidly frozen by subjecting a thin stream of molten metal to the impact of a high-energy jet of gas or liquid. Air, nitrogen and argon are the commonly used gases, and water is used in the production of iron powder.

Reduction of the metal from oxide is another method of making iron powder. A selected ore is crushed, mixed with carbon and passed through a continuous furnace where the reduction takes place, leaving a cake of sponge iron, which is then crushed and sieved, to produce the metal powder. Copper, some refractory metals and, in at least one plant, iron, are made as powder by reducing the oxides with hydrogen. Many metals can be deposited by electrolysing them to form a spongy or powdery condition. Copper is the main metal to be powdered in this way, but chromium and manganese powders are also produced.

If necessary the metal powder is then blended with the material with which it will be associated. Tungsten carbide blended with cobalt for cutting tools was the first of this class of product and still represents the major part of the market. Other carbides, nitrides and borides are being used in increasing quantities. In another type of blending, friction materials for brake linings and clutch facings are embedded in copper or other materials.

Next the powder, mixed with a small quantity of lubricant, is compacted at room temperature in tungsten carbide or high-speed steel dies, shaped to the finished form of the component. The dies are mounted on machines which can provide a pressure of 150–1000 N/mm², depending on the metal being treated and the size of the part. This compresses the powder into the die cavity and binds it together. The die is opened and the moulding automatically ejected, then the lubricant is removed. The compacted component is not at all strong at this stage, but it can be carefully handled.

The final process, sintering, is a key part of the operation, giving the required strength to the friable compacted powder. The operation is generally carried out under a protective atmosphere at temperatures of between 60 and 90 per cent of the melting point of the alloy or metal. Gas or oil furnace heating can be used up to 1000°C; higher temperatures are obtained by electrical resistance heating. For maximum production it is desirable to operate the furnace on a continuous basis and the use of mesh belt conveyors is general practice. Carbides for tool tips are sintered, first in a reducing atmosphere, generally hydrogen, then in vacuum or in argon.

About two thirds of the products of ferrous powder metallurgy are required for automotive components. These include sprockets, timing pulleys, components of starter motors, lock parts, bushes, linings for brakes and clutches. Domestic electrical appliances, business machines, telecommunication equipment, cameras, and DIY products make up about a tenth of the total. In Japan an enormous business has been created for sintered magnetic materials, using the new alloys containing one or more of the rare earth metals; this market alone accounts for about 100 000 tonnes per annum.

Although most sintered products come within a weight range of 2 grams to 3 kilograms, there have been some exceptions – up to 16 kilograms. Initial tool costs are fairly high, so production runs are justified only if quantities in excess of 10 000 items per annum are required. Many sintered products are ordered by the million. Design limitations are somewhat similar to those for diecasting because the parts are compacted in permanent metal dies: undercuts should be avoided and, if possible, thin walls and sharp corners are 'designed out'. The speed of production depends very much on the size of the component and the number of impressions in the die. In one case the die contained 30 impressions of a small item and about 4500 mouldings per hour were produced. For single impression dies, outputs of several hundred per hour can be expected.

In recent years several processes have been developed to make sintered metal products more economical and to achieve a closer approach to the finished article. In one process the powder is compressed in a die at a high temperature, thus combining the pressing and sintering into one operation. This has the advantage over cold pressing of requiring lower pressures. When the temperature required is too high for metal dies carbon or graphite can be used.

Another method is known as isostatic pressing in which the metal powder is packed in a flexible container which is then subjected to hydraulic pressure. There are two varieties of the process. In cold isostatic pressing the powder is encapsulated in a container, of rubber or plastic, which is then immersed in a liquid, usually water, which is pumped to a high pressure of about 4000 atmospheres. The powder is compacted from all directions and uniform density can be achieved. No lubricant is needed. Nowadays this process is automated, giving high production rates.

Hot isostatic pressing is also being used on a large scale. The container is usually of metal because the process takes place at an elevated temperature. Pressure is applied by argon and needs to be only one quarter of that used in the cold process. Hot pressing is used for the moulding of hard metals, such as those required for extrusion or other dies. It is also applied in the production of billets of superalloys, high speed steels and titanium, where the integrity of the material is the prime consideration.

The strength of metals when powdered, pressed and sintered can be increased, and greater accuracy obtained, by coining. In this operation, the piece is forced into a shaped die which is slightly smaller than the piece itself. Another development is impregnation or infiltration. A pressed piece of iron is sintered in such a way that it is extremely porous, and then it is placed into molten copper until all the pores are filled.

SINTERED CUTTING TOOLS

For years metallurgists and engineers sought tools that would be superior to high-speed steels for cutting and machining metals. It was found that when tungsten powder was heated with carbon to a temperature of about 1500°C a compound, tungsten carbide, was formed; this is exceedingly hard and very suitable for making cutting tool tips, now used in many millions of appliances. The tips are made by mixing tungsten carbide powder with 6 to 10 per cent cobalt. The mixture is pressed into blocks and then heated in hydrogen to a temperature of about 1000°C; this pre-sintering makes a product which, though hard, can still be machined and ground to the required shape and size. The tool tips are then given a final sintering at about 1400°C, to become much harder than the hardest steel. The tips are brazed or mechanically locked on the end of steel shanks. The great hardness of tungsten carbide has produced drills that are able, for example, to drill holes in glass. Tungsten carbide is used for making dies or die inserts for such processes as extrusion and wire drawing, and for the compacting tools that produce other powder metal components. A variety of intermetallic compounds can be added to give even greater hardness. Cobalt has already been mentioned; titanium carbide or tantalum carbide are also added.

So called 'oil-less' or 'self-lubricating' bearings are made by sintering a mixture of copper and tin powders with graphite. They are then soaked in oil, and no oil holes or grooves are required since oil will soak through such a bearing as though it were a sponge. In some cases, the amount of oil held in the bearing is enough to last the life of the machine, and some are so porous that they could be used as wicks in oil lamps. Self-lubricating bearings are used in the automobile industry, and in many domestic articles such as washing machines, vacuum cleaners, and electric clocks.

Many of the so-called 'new metals', including tantalum, molybdenum, and niobium, as well as some of the older metals, such as nickel, copper, cobalt and iron, can be extracted from their ores to yield the metal in the form of powder. The cost of such powders is not higher than solid metal and may in fact be lower. Hence there is an incentive to fabricate by powder-metallurgical methods; and 'ingots' of molybdenum and of nickel, weighing a tonne or more, have been made by pressing and sintering.

Powder metallurgy is a comparatively young and virile industry and after a period in the 1950s, when it was finding its feet, it has been moving ahead with considerable verve. In addition to the obvious advantages of producing a finished component that does not require further machining it offers some advantages relevant to today's problems. Process scrap is very small, amounting to only about a few per cent. This can be compared with the foundry industry where the surplus metal – the runners and risers – may amount to a third or more of the weight of the finished component.

23

METALS AND NUCLEAR ENERGY

Atoms are very small and very numerous; in a piece of metal the size of a pin head there are many more atoms than the total of all the letters in every book that Penguin Books has ever published. Following the work of nineteenth-century chemists, including John Dalton and Henri Lavoisier, the relative atomic weights were calculated, compared with unity for hydrogen, but it was not until the twentieth century that their actual weights and sizes were discovered. Before the First World War, Ernest Rutherford in Britain and Niels Bohr in Denmark suggested that each atom contains a central nucleus which is positively charged; negatively charged electrons rotate round the nucleus but are prevented from uniting with it by their speed in orbit, just as the earth is prevented from falling into the sun. The atom of hydrogen has one electron circling round a nucleus of one positive charge, the proton. Helium has two electrons round a nucleus containing two protons; lithium three, iron twenty-six and uranium ninety-two. The number of protons in the atom of an element is known as its atomic number.

Electrons have only about one eighteen-hundredth the mass of protons; they behave as though they move around their nucleus in one or more rings. Hydrogen and helium have only one ring; lithium's three electrons arrange in two rings; aluminium has three rings for its thirteen electrons; iron, nickel, copper and zinc have their electrons in four rings; some of the heavy metals including silver and molybdenum have five rings; gold and the rare earth metals have six rings. Uranium and radium have seven rings – the most ever observed. The outer-ring electrons are principally responsible for the chemical properties of each element.

ISOTOPES

When the relative atomic weights had been calculated from chemical experiments it was at first surprising that they were not whole integers; for example, chlorine was found to have an atomic weight of $35\frac{1}{2}$. The explanation came in 1913, from a British scientist, Frederick Soddy, who deduced the existence of 'isotopes', which are atoms of the same element, differing in atomic weight but which have identical chemical properties and identical physical properties, except those determined by the mass of isotopes. In 1919 Frederick Aston was able to prove that about three quarters of chlorine has an atomic weight of 35, while the other quarter has an atomic weight of 37; this mixture gives the actual atomic weight $35\frac{1}{2}$ mentioned above. Since the chemical properties of the two isotopes of chlorine are identical, they cannot be separated by any chemical process, but Aston isolated them in minute amounts by the mass spectrograph. Under an appropriate electromagnetic field, the two isotopes can be made to follow slightly different paths, because of their differing masses; they can then be separated and identified, much in the same way as two rockets of differing weights fired off with a similar thrust will follow differing trajectories.

Although most elements are a mixture of two or more isotopes and sometimes (tin for example) as many as ten, one isotope is generally present in a preponderant amount. Thus oxygen contains 99·80 per cent of oxygen 16, 0·03 per cent of oxygen 17 and 0·17 per cent of oxygen 18. Even hydrogen was found to contain 0·02 per cent of an isotope with atomic weight 2, which became known as 'heavy hydrogen' or 'deuterium'. 'Heavy water', formed by uniting heavy hydrogen with oxygen, has important uses in nuclear energy research, and in some nuclear power stations. It is usually given the symbol D_2O, to distinguish it from H_2O.

Having proved the existence of isotopes, the question was 'How could an element change its mass without changing its chemical properties?'. The answers to this and other mysteries were obtained after a suggestion, by Sir James Chadwick, that the nucleus of the atom is complex in structure. Each element has a characteristic number of protons, the positively charged particles; however the nucleus also contains neutrons, which are uncharged particles having very nearly the same weight as a proton. Electrons, equal in number to the protons, and each with a negative electric charge counterbalancing the positive charge of a proton, circulate round the nucleus in one or more rings as described above. The presence of the neutrons does not affect the chemical properties of the element but only changes the mass of the nucleus. In a given element all the isotopes have the same number of electrons, which provide its characteristic chemical properties. They have the same number of protons in the nucleus but each isotope has an individual number of neutrons. It was discovered later that different isotopes of some metals, including uranium, possessed very individual properties in their radioactive behaviour and in their response to 'fission'.

SPLITTING THE ATOM

Ernest Rutherford and his co-workers realized that it should be possible to break away some protons from the nucleus of an atom. The most suitable method then available was to bombard the nucleus with a stream of the 'alpha particles' which are spontaneously thrown off from radioactive elements such as radium. They are the nuclei of helium atoms which have lost their circulating electrons and so have a nett positive charge. By directing a beam of these alpha particles on to thin layers of various elements, Rutherford was able to witness scintillations on a fluorescent screen, indicating that a nucleus had been hit, but these collisions were only achieved atom by atom and thus the mass affected was infinitesimal. In the course of these experiments it was suggested that if the atom could be 'split', a valuable source of energy might be obtained; indeed the 'man in the street' in the nineteen-thirties was confident that the splitting of the atom would solve all our energy problems. But unfortunately it required much more energy to break a small fragment away from an atom than the energy given out in the experiment.

The problem was twofold; the targets were very small; if a nucleus can be said to have a 'diameter' it is only one ten-thousandth the size of a 'diameter' of an atom. Secondly, a positively charged particle used as a projectile would have difficulty in hitting its minute target, because positive charges repel each other. When Ernest Rutherford was pressed to give a prophecy about the atom as a source of energy he remarked, 'Anyone who looks for a source of power in the transformation of the atom is talking moonshine'.

Chadwick's discovery of the uncharged neutrons provided a type of particle which would not be repelled by the proton nucleus, but the problem remained depressingly elusive because of the atom-by-atom approach that was necessary. What was needed, but what did not then exist, was a method by which when one atom was split, the disintegration separated further neutrons, thus causing what was later to become known as a 'chain reaction'. If for example two neutrons could be given off, they might cause two other atoms to be split, which would yield four neutrons, then leading to 8, 16, 32, 64, 128, and so on in rapidly increasing amounts.

During pioneering experiments in neutron bombardment, many materials were changed, in minute quantity, to new isotopes or new elements.

The mechanism of this process may be understood by considering an atom of carbon, which consists of six protons and six neutrons in the nucleus and six electrons circulating. If the carbon atom is bombarded by neutrons, it is converted into an isotope with six protons and *seven* neutrons, still with six electrons. This is a stable isotope 'carbon 13' which is contained in natural carbon to the extent of about one per cent.* The specimen of carbon bombarded by the neutrons would be made to contain more atoms of carbon 13 than occur in natural carbon.

On the other hand, if we bombarded sodium, which has only one stable isotope and in which all the atoms contain 11 electrons, 11 protons and 12 neutrons, we would produce an unstable form of sodium with 13 neutrons. One neutron is surplus and cannot be held in the nucleus; it can change from the unstable condition by acquiring a positive charge which turns it into an extra proton. Since the neutron is an electrically neutral particle, it achieves this transformation by giving out a negative charge (that is an electron). This brings the atom to a condition where it now consists of 12 protons (an increase of one), 12 neutrons (same as before), and therefore the atom is no longer that of sodium, but has been converted to the next element, magnesium, which has 12 electrons.

Realizing that neutron bombardment could convert an element into one with a higher atomic number, the Italian physicist Enrico Fermi had the idea of creating new elements by bombarding uranium, which, with 92 protons, was the largest known atom. The metal is composed principally of uranium 238, but less than one per cent of its mass derives from the isotope uranium 235. Following the principle described above for converting sodium into magnesium, Fermi succeeded in proving that he had converted part of the uranium, first into neptunium, with 93 protons, and this in turn into plutonium, with 94 protons; both these are synthetic elements.

THE USE OF URANIUM FOR NUCLEAR ENERGY

In 1938 the German physicists Otto Hahn and Fritz Strassmann were studying the bombardment of uranium by neutrons and repeated Fermi's experiments. At first they thought they had produced only 'transuranic elements' like plutonium, but on again repeating the experiments they discovered that other, lighter, elements had also been formed, including the metal barium and an inert gas, probably krypton, both in such minute amounts that their existence could barely be identified. An Austrian scientist, Dr Lise Meitner, and her nephew Otto

---

* Natural carbon also contains an infinitesimal amount of carbon 14, used in determining the age of antiquities. It has eight neutrons, and is radioactive.

Frisch interpreted these unexpected results as indicating that the nucleus of the uranium 235 isotope had been split into several fragments with atomic weights ranging from about one third to two thirds that of uranium, coupled with the production of several secondary neutrons. In contrast with this, Fermi's experiments had only revealed that the atoms of uranium 238, comprising the major part of the metal, had been changed into slightly larger atoms, neptunium and plutonium. Hahn and Strassmann's momentous findings, as interpreted by Lise Meitner, showed that they had also achieved 'fission' of the 235 isotope, which behaved in a different manner from the 238 isotope.

If the formation of the fragments were the only result of fission of the uranium atom, the process would be of no greater value in the provision of large-scale energy than Rutherford's early experiments. But the fission of uranium 235 also releases two or three neutrons. Thus we reach the position envisaged on page 255 in which the splitting of one atom provides two or more projectiles well suited to split more atoms, initiating a chain reaction capable of continuous self-activation, with the release of energy. The weight of the nucleus of uranium 235 is about one thousandth part heavier than the sum of the weights of the fission fragments. Einstein's equation states that $E = MC^2$, where E is a measure of the amount of energy liberated when a mass M is annihilated; C is the velocity of light. If the whole matter of only one gram of a substance could be destroyed and changed into energy, over two hundred million kilowatt hours would be produced.

Only a few days before Hitler invaded Holland, Belgium and France, this discovery was reported in the newspapers; immediately afterwards, for reasons which are now well apparent, further references to the fission of uranium 235 were put under the seal of security until after Hiroshima, over five years later.

THE ATOMIC BOMB

The amount of the 235 isotope contained in the small sample of uranium bombarded by Hahn and Strassmann was minute; the neutrons were fast-moving and practically all of them escaped, so that a chain reaction did not take place. If, however, a mass of uranium 235 could be obtained, having a diameter of about 120 mm (a sphere of this size would weigh about 16 kg), this would exceed what is known as the 'critical mass'; sufficient neutrons would be given enough chance to collide with more and more uranium 235 atoms without escaping. Under these conditions a chain reaction could occur, thus annihilating a fraction of the mass and instantaneously evolving a vast amount of energy. Simplifying as we must, the process required is to make two portions of uranium 235, each of which exceeds half the critical mass. The two portions must be separated until the required moment for explosion, when they are rammed together. Immediately they touch, the critical mass is exceeded and the chain reaction initiated by the neutrons can take place.

The separation of uranium 235 is an operation of prodigious complexity and cost. The chemical properties of the isotopes of an element are identical, so they have to be separated by physical methods, relying on the slightly different atomic weights. One method is to extend the use of the mass spectrograph, referred to on page 254. Another method is to make the compound uranium hexafluoride, which when warmed becomes gaseous. Those molecules of the gas which contain the 235 isotope are a little lighter than the major part, containing uranium 238, and, in the gaseous form, move about or diffuse at a slightly greater rate than the heavier ones, the difference being about one per cent.

The gaseous uranium hexafluoride is made to diffuse through very fine pores in a series of multiple diaphragms. The faster moving molecules pass through somewhat more readily than the heavier ones so, as the gas diffuses through each diaphragm, the proportion of the lighter molecules is slightly increased. The process repeated over and over again gradually separates the proportion of uranium hexafluoride containing the 235 isotope. The scale on which this process is necessary can be indicated by remembering that the plant at Oak Ridge, Tennessee, U.S.A., employed 75 000 people in a number of immense buildings, spread over seventy square miles, with a maximum output of only 3 kg of uranium 235 per day, and consumed a vast amount of electric power in the process, particularly because an almost complete separation of the two isotopes was required for the atomic bomb.

Uranium hexafluoride is highly corrosive and will destroy many of the normal materials of which chemical apparatus is made. The compound is itself liable to decomposition, particularly if moisture is present, and, unless stringent precautions are taken, will change to a solid material deposited on the walls of the diaphragm membranes, evolving the poisonous fluorine and hydrofluoric acid. Like some other abominable things, it is given a friendly nickname – 'Hex'.

THE GAS CENTRIFUGE

Modern nuclear reactors require the proportion of the 235 isotope to be

Metals and Nuclear Energy 259

increased from its natural level of 0·7 per cent to about 3 per cent. Until the 1960s, the gaseous diffusion plants described above were the principal methods of enriching uranium with the 235 isotope. However, as far back as 1940 the isotope had been separated experimentally by centrifuges but at that time the technology of high speed rotating machinery was not equal to the task, so the centrifugal method was abandoned in favour of the gaseous diffusion process. Improvements in gas centrifuges were being made continually and in 1960 the U.S. Atomic Energy Commission authorized a development programme which led to the establishment of enrichment plants in which the amount of electric power, though still enormous, is less than one tenth that required in the gaseous diffusion of 'Hex'. In 1970 Britain, the Netherlands and West Germany began a joint project, to develop the gas centrifuge process. A pilot plant was built in Capenhurst while the Dutch and German plants were built at Almelo.

Uranium hexafluoride is heated to 80°C to convert it into gas, which is fed into a series of vacuum tanks each containing a rotor about 1 m long. When the rotors are spun rapidly the heavier molecules, based on the 238 isotope, move nearer towards the circumference than the lighter 235 isotopes. In modern counter-current centrifuges, 'scoops' inside the drum make it possible to feed the gas enriched in 235 and the depleted gas to separate exits at the top. Each rotor enriches one tenth of a gram of Hex per second; the process is repeated over and over again in cascades of several thousand rotors.

The peripheral speed of each rotor needs to be at least 400 metres per second – much faster than the speed of Richard Noble's car that broke the land speed record. The basic object in designing a centrifuge is to provide the fastest and longest rotor possible, with maximum life and minimum cost. The rotors must be of very strong material. Stainless steel and some aluminium alloys will withstand 400 m per second; titanium alloys up to 440 m per second and maraging steels, discussed on page 192, up to 525 m per second. For still higher speeds a fibreglass/polyester composite has been developed in Europe.

Thousands of miles of piping and at least 100 000 welds are required in a centrifuge plant, while handling the corrosive and poisonous Hex adds many more problems. The scale is enormous: one enrichment plant being constructed in Ohio has eight process buildings, each the size of four football fields. The number of centrifuges for the world's requirements of enriched uranium will entail several million centrifuges and several billion dollars of capital investment.

CONTROLLED FISSION FOR NUCLEAR ENERGY

When scientists began to turn their attention from the costly production of atomic bombs to the possibilities of harnessing nuclear power for peaceful purposes, it was realized that, for economic reasons, natural uranium, or the metal only slightly enriched to about 3 per cent of the 235 isotope, should be used.

Heat is created by the fission of the 235 isotope which forms $0 \cdot 7$ per cent of the mass of natural uranium. This can be made to split if hit by slow neutrons, dividing the 235 isotope into fragments, releasing heat energy and two or three fast-moving neutrons. These must be slowed down by what is called a moderator. Ordinary water is an effective moderator but unfortunately it also captures some of the neutrons, so there are fewer available to cause fission of the uranium 235. Graphite and heavy water capture far fewer neutrons so, if either of them is used as moderator, controlled fission can take place with the $0 \cdot 7$ per cent of the 235 isotope that is present in natural uranium. With ordinary water as moderator the $0 \cdot 7$ per cent is not sufficient to allow the chain reaction to continue; about 3 per cent uranium 235 is required and the uranium has to be enriched, using methods such as those described on the previous pages. Enrichment plants are very costly to build and use immense amounts of energy, so a graphite moderator with natural uranium was most suitable and economic for the first nuclear power stations.

It is worth reminding ourselves that in the atomic bomb, which was made of over 90 per cent uranium 235, fast neutrons initiated the fission. So long as the mass of U 235 was less than critical, the fast neutrons would escape before having a chance to cause fission. When the critical mass was exceeded the fast neutrons became so thoroughly trapped that the chain reaction immediately took place. Fast neutrons travel at speeds of many thousands of kilometres per second. In the controlled conditions of the nuclear reactor slow neutrons are harnessed, having speeds of only a few kilometres per second. Such neutrons are commonly called thermal neutrons. Some of these hit and split uranium 235 atoms, thus bringing about a chain reaction. However, unlike the condition in the atomic bomb, this is a controlled chain reaction, regulated by materials which can absorb neutrons.

NUCLEAR POWER STATIONS

In 1945 the British Government decided that the nuclear research centres at Harwell and at Risley should investigate the use of nuclear energy for power. By 1950 the possibilities of producing electricity in this way had been examined, and appeared likely to be economic. Three years later, construction of the Calder Hall nuclear power station was begun; in the following year, the U.K. Atomic Energy Authority was set up. On 17 October 1956 Calder Hall, the world's first full-sized nuclear power station, was opened by the Queen. This station was the first of several, all operating on similar principles. They are known as gas-cooled reactors. *Fig. 50* is a schematic drawing to illustrate the principles of a gas-cooled reactor. The information given below is specific to Calder Hall, in order to give some idea of the scale of these reactors, and to indicate the function of the metals that are used, but it will be realized that in any series of nuclear reactors there are developments, improvements and changes in size.

The station consists of four reactors housed in cylindrical steel pressure vessels 20 m high. Each reactor shell surrounds a large graphite construction weighing 1100 tonnes, made up in the form of 14 000 bricks and 30 500 tile-shaped components, enclosing about 10 000 rods of natural uranium (that is, with only 0·7 per cent of the 235 isotope). Each rod is about a metre long and 30 mm in diameter; the total weight of uranium is about 120 tonnes in each reactor.

The uranium is sealed in metal 'cans' to contain the waste products of fission. Each can is finned, to assist the transfer of heat, and it seals the uranium to prevent it from being oxidized by the carbon dioxide gas which transfers heat from the reactor to the heat exchangers. The mundane word 'can' hardly does justice to an object which is difficult and expensive to make, and which possesses a kind of beauty, as will be seen from *Plate 20* which shows a selection of cans for containing the fuel elements. These cans are made of a magnesium-rich alloy known as Magnox, which is discussed later.

The speed of the neutrons must be controlled to prevent excessive heat being developed; this is done by selecting a material which can entrap neutrons; the control rods made of it can be adjusted to retain a steady 'neutron cross-section' in the reactor. During normal operation the control rods are moved in a very small travel within the reactor core by electric motors; this movement is finely controlled to fractions of a millimetre. The control rods are suspended electro-magnetically so that, in the event of electrical failure, they would drop fully home, their full length of 7 to 10 metres then being entirely in the reactor core; this would immediately stop the chain reaction. In the Calder Hall reactor there are 48 control rods, made of a steel containing boron.

The heat of the reactor is taken away by forcing carbon dioxide under pressure through the system. The gas enters the reactor at 140°C and is heated to about 340°C. One of the many facts that surprise a visitor is

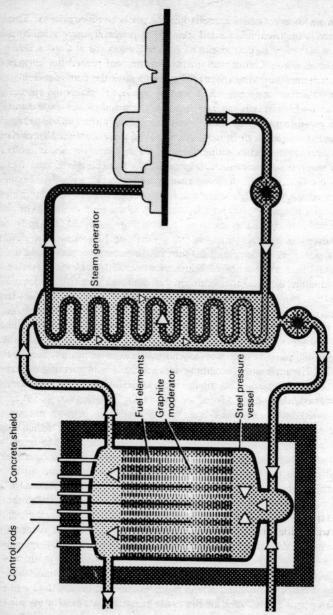

Control rods

Concrete shield

Fuel elements

Graphite moderator

Steel pressure vessel

Steam generator

Fig. 50. Basic gas-cooled reactor (Magnox) (Courtesy U.K. Atomic Energy Authority)

that as much as a tonne of carbon dioxide per second circulates. The reactor core is enclosed in a welded steel pressure vessel surrounded by a concrete shield; over 10 000 tonnes of concrete was used at Calder Hall. Having heated the carbon dioxide in the reactor, and passed the hot gas to a heat exchanger, the remainder of the power station is conventional.

Eleven of these stations, often called 'Magnox' referring to the magnesium alloy of which the cans are made, proved to be prolific and economic power-producers. Other stations are larger than Calder Hall. For example, compared with the 1100 tonnes of graphite there, Sizewell contains over 2000 tonnes while Wylfa in Anglesey has about 3700 tonnes. The pressure vessels at Calder Hall and Sizewell are of steel 100 mm thick but that at Wylfa is of concrete $2\frac{1}{2}$ metres thick.

ADVANCED GAS-COOLED REACTORS

While Calder Hall was getting into its stride, the prototype of a new concept, known as the Advanced Gas-cooled Reactor (abbreviated to AGR), was being constructed at Windscale, near Calder Hall. This uses uranium dioxide as fuel, the uranium having been slightly enriched, giving a longer life to the fuel and allowing the carbon dioxide temperatures to be about 200°C higher than in gas-cooled reactors. As Magnox cans are not suitable to endure this greater heat, the uranium oxide pellets are inserted into stainless steel cans. The fuel elements are referred to as 'pins', which are assembled into clusters; the reactor contains 200 channels each containing 4 clusters of 9 pins, making a total of 7200. The use of uranium dioxide was an important improvement; modern stations do not use fissile elements in the metallic form. *Fig. 51* illustrates the principle of an AGR.

Fig. 52 shows some details of a fuel element in an AGR. The fuel can, guide tube, support grid and centre brace are of stainless steel, containing an added stabilizer element, which is titanium for the centre brace and niobium for all the others. The graphite of the sleeve is of a special nuclear grade, having low neutron absorption. *Plate 21* shows the immense equipment which is required to load the fuel elements into the AGR at Hinkley Point B nuclear power station.

There were initial problems with corrosion of boiler materials, vibration and noise within the reactors. Also it became necessary to develop special thermal insulation and methods of inspecting inaccessible components. The AGR stations suffered from bad environmental publicity and some of them took a long time to get going, but it is expected that in their pre-stressed concrete vessels, AGRs will prove to be satisfactory, and ultra-safe.

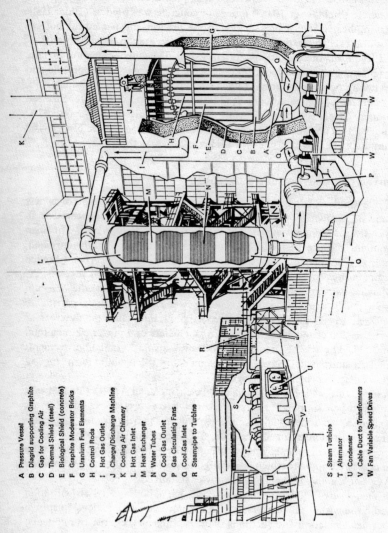

A Pressure Vessel
B Diagrid supporting Graphite
C Gap for Cooling Air
D Thermal Shield (steel)
E Biological Shield (concrete)
F Graphite Moderator Bricks
G Uranium Fuel Elements
H Control Rods
I Hot Gas Outlet
J Charge/Discharge Machine
K Cooling Air Chimney
L Hot Gas Inlet
M Heat Exchanger
N Water Tubes
O Cool Gas Outlet
P Gas Circulating Fans
Q Cool Gas Inlet
R Steampipe to Turbine

S Steam Turbine
T Alternator
U Condenser
V Cable Duct to Transformers
W Fan Variable Speed Drives

Fig. 51. Advanced gas-cooled reactor (*Courtesy U.K. Atomic Energy Authority*)

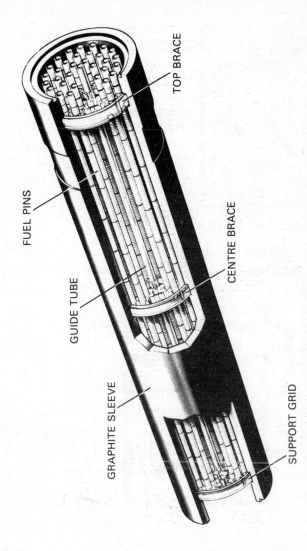

TOP BRACE

FUEL PINS

CENTRE BRACE

GUIDE TUBE

GRAPHITE SLEEVE

SUPPORT GRID

Fig. 52. Advanced gas-cooled reactor fuel element (*Courtesy British Nuclear Fuels plc*)

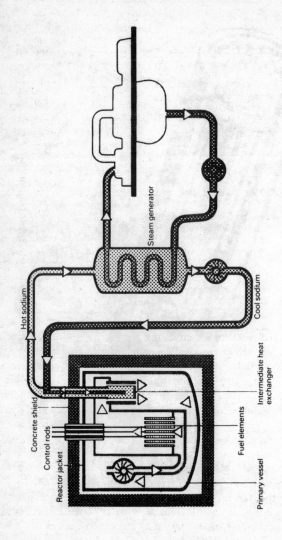

Fig. 53. Sodium-cooled fast reactor (*Courtesy U.K. Atomic Energy Authority*)

Table 20

	Magnox gas-cooled reactor	Advanced gas-cooled reactor	Fast breeder reactor
TYPICAL SITE	Sizewell	Hinkley Point 'B'	Dounreay
NUCLEAR FUEL	Natural uranium	Enriched uranium oxide	Mixed enriched uranium and plutonium oxides
MODERATOR	Graphite	Graphite	None
FUEL CANS	Magnox	Stainless steel	Stainless steel
CONTROL RODS	Boron steel	Boron steel	Tantalum or boron carbide (under investigation)
COOLANT	Carbon dioxide	Carbon dioxide	Liquid sodium
SIZE OF REACTOR (*cubic metres*)	1320	550	$4\frac{1}{2}$

THE FAST BREEDER REACTOR

A more advanced reactor, known as the fast breeder, illustrated in *Fig. 53*, uses a different concept in which no moderator is required. A mixture of oxides of uranium and plutonium is clad in stainless steel and made up into compact hexagonal fuel element assemblies. By including a blanket of uranium waste product from enrichment plants, that material is converted, by neutron capture, first into neptunium and then to plutonium 239 which is a fissionable isotope; the process is known as breeding. The considerable amount of heat evolved is removed by molten sodium which, although it has some unpleasant characteristics, is relatively cheap and, having a high thermal conductivity, it is an excellent material for transferring heat from the reactor core. At Dounreay in the far north of Scotland an experimental plant was producing about 60 megawatts from 1959 to 1977. In the meantime a pilot station was built, also at Dounreay, which has paved the way for the eventual development of this form of reactor.

Table 20 shows some details of the three types of reactors described above. The comparison in size of reactor cores is interesting – the gas-cooled reactor the size of a medium-sized cinema, the advanced gas-cooled reactor a small bungalow, and the core of the fast breeder reactor not much bigger than a dustbin. The fast reactor produces about forty times more electricity per tonne of uranium than the gas-cooled reactors and it is more efficient in using our nuclear resources because it also creates plutonium from uranium waste.

OTHER NUCLEAR REACTORS

The Pressurized Water Reactor (PWR) was developed in America and the U.S.S.R. for marine propulsion and is now the most widely used type of reactor in the world. More than twenty countries have PWRs. Uranium oxide with the metal enriched to 3·2 per cent of the 235 isotope is clad in zirconium alloy; the pins are arranged in clusters of as many as 264 rods, held together in a bundle, weighing about 0·75 tonnes. The Boiling Water Reactor (BWR) uses uranium oxide in zirconium alloy cans; the moderator is ordinary water. The Steam Generating Heavy Water Reactor (SGHWR) also uses uranium oxide in zirconium alloy cans but the moderator is heavy water.

Many hitherto unexplored problems in nuclear engineering and metallurgy can be solved by tests in universities and nuclear research establishments which are equipped for the development of new materials and processes, often leading to useful discoveries with commercial and export potential. For example, silicon chips are doped with phosphorus by neutron irradiation at Harwell in the U.K. Silicon contains an isotope which can be transformed into phosphorus by a nuclear reaction with the neutrons; because that isotope is distributed absolutely uniformly, the distribution of the phosphorus in the remaining silicon is accordingly regular – more so than when it is doped chemically. Large single crystals of silicon are sliced into chips and doped by neutron irradiation, providing chips for new generation equipment, ranging from ultra-fast computers to miniature watches.

THE ROLE OF THE METALLURGIST IN NUCLEAR ENERGY

All the systems for producing power from nuclear fission pose new problems for the metallurgist: the fissile materials themselves, the canning alloys, the control rods and the special problems of corrosion under the extraordinary conditions inside the reactor. Some of the metals now used quite extensively were merely laboratory curiosities forty years ago and it has been necessary to discover the best ways of extracting them from their ores, to cast, forge and join them and to investigate their physical and mechanical properties.

Metallurgically, uranium has a number of peculiarities. It oxidizes rapidly in air at 200°C and above; its atomic lattice structure at room temperature is complex, and each crystal of the metal expands non-uniformly when heated. This leads to internal stresses when the metal undergoes heating and cooling. 12–15 per cent of the fission product consists of the inert gases xenon and krypton, which tend to collect in fine bubbles and exert internal pressure which can lead to large volume changes. Another distortion problem connected with the properties of uranium is growth due to the effect of radiation. One direction in the uranium crystal gets longer while another contracts. This appears at its worst in rolled uranium and can cause the bar to double its length.

As mentioned above, the fuel used in most modern reactors is uranium dioxide and is made by sintering powder at approximately 1650°C in an inert or slightly reducing atmosphere. The finished product has about 95 per cent of the possible theoretical density. During the early stages in a reactor further sintering occurs and the fuel pellets shrink a little. Although this shrinkage is negligible on the diameter of the pellets, the change in length may add up to one or two centimetres. A problem arises if part of the stack of pellets gets stuck in the tube, leaving a gap between pellets. Sudden release means that the reactivity of the reactor is changed. This is one more reason for great accuracy in manufacturing tolerances.

The nuclear industry, both in reactors and processing plants, uses a very large amount of stainless steel. During the past forty years, with mutual cooperation, vast strides have been made in the quality of stainless steel to meet the demands of the nuclear industry. This cooperation has also led to new methods of making thin steel tubes and significant developments in welding and inspection techniques.

CANNING MATERIALS

The choice of the correct material for the cans required in a nuclear power station depends principally on the type of reactor and especially on the working temperature inside the reactor, but a number of other technical problems are involved. The metal must not capture too many neutrons, otherwise the efficiency of the reactor will be lowered. The cans must be able to withstand the cooling-gas environment at the operating temperature. The canning material should conduct heat reasonably well and must withstand the temperature at which the reaction is controlled, up to 800°C in advanced gas-cooled reactors. It must be capable of being formed into thin-walled containers, with fins to aid dissipation of heat, and it must be sealed effectively after the uranium oxide fuel has been inserted.

The canning material must be sufficiently ductile to allow for deformation and distortion in the reactor. It must have enough strength to support its own weight under the effect of heat and radiation and under any stresses caused by warping of the fuel elements. Once a nuclear reactor has achieved its controlled chain reaction, by-products are formed: plutonium and extremely radioactive isotopes of a wide range of elements. Some of these are gases, krypton and xenon, and space has to be left inside the cladding to avoid high pressures. Other fission products, such as iodine and caesium, can corrode the inside of the cladding. In fast reactors as much as 20 per cent of the uranium and plutonium is changed into fission products. There is continuous monitoring of the solidity of cans in the reactor so that any leak can be detected quickly and the offending can removed.

When the uranium or its oxide are 'spent', the contents of the can must be capable of being dissolved in chemical agents, by remote control, so that uranium can be reclaimed for further use and plutonium recovered. The cans themselves have to be stored in silos because, by the time their job in the reactor is finished, they are highly reactive and will remain so for many hundreds of years.

Magnesium was a suitable can material for the first design of gas-cooled reactors, as it complied to a greater or lesser extent with the requirements listed above. A magnesium alloy, containing about 0·8 per cent aluminium and 0·01 per cent beryllium, known as Magnox A L, was used to sheathe the uranium fuel in those reactors. The small additions to the magnesium ensure that a more tenacious and protective oxide skin is formed than would be obtained with the pure metal.

As processes which evolve greater heat are developed, differing canning materials are required. Magnox would not be suitable at temperatures higher than those existing in the first design of gas-cooled reactor. At first it was thought that beryllium might be an ideal material for cans in advanced gas-cooled reactors but metallurgical problems, as yet unsolved, caused the large-scale production of beryllium to be suspended. Stainless steel has the advantages of resisting the heat and not warping, but it absorbs more neutrons than beryllium and has a lower heat-conductivity. To get optimum results, a stainless steel can has to be only about a third of a millimetre in wall section thickness. Advanced gas-cooled reactors are now using a steel with 20 per cent chromium, 25 per cent nickel and about 1 per cent niobium.

SOME MISCELLANEOUS PROBLEMS

The control rods have to be mechanically strong and must have a very high absorption of neutrons. In gas-cooled Magnox reactors, the control rods are of steel containing 3–4 per cent boron, encased in stainless steel sheaths. The advanced gas-cooled reactors use two types of control rods, either boron steel inserts in stainless steel sheaths or simply rods of stainless steel. Recently a control rod alloy containing 80 per cent silver, 15 per cent indium and 5 per cent cadmium has been developed, mainly for the various types of water reactors mentioned on page 268, where optimum resistance to the corrosion of the coolant is required. For more compact or more sophisticated reactors, other materials, sometimes exceedingly costly ones, are chosen for their great ability to absorb neutrons. Since hafnium became more available than before, it is being specified for control rods of compact reactors. By far the best absorber of neutrons is the metal gadolinium (see page 218), but supplies are so scarce that its use is limited to special reactors which require control rods of extraordinarily high efficiency.

Early gas-cooled reactors embodied steel pressure vessels but now pre-stressed concrete is used. Pressurized water reactors operate at such high pressures that steel vessels are essential, often with a thickness over 200 mm; the welding of this plate and its heat treatment to relieve internal stresses are precise and important operations. Pressure vessels have also been the subject of much debate, especially concerning the cracking of the steel walls during service. The laboratories of the Atomic Energy Authority have set up special equipment to test the very thick sections which are used.

Special attention must be given to 'compatibility', which involves similar problems to sacrificial corrosion mentioned on page 232. The metals in the nuclear power unit must not deteriorate through contact with other materials. For example, stainless steel components in a fast breeder reactor will have to 'sit' in molten sodium and be exposed to fast neutrons for thirty years, all in addition to the thermal stresses which must be endured.

The neutrons in a fast reactor have enough power to knock metal atoms out of their place in the crystal structure. In fact every atom of the stainless steel components inside the reactor gets knocked out of position tens of times during its operational lifetime. The minute holes left behind the knocked-out atoms can join together and eventually the strain causes a few per cent of swelling. The most serious effect is when one side of a component, for example a tube, gets hit by more neutrons than the other side, causing the tube to bow. A way round this is to turn the tube at regular intervals.

Like a blast furnace or a rolling mill the nuclear power station has to be kept running with a maximum of economy, output and safety. Every change in conditions may introduce the possibility of some problem arising and the most scrupulous precautions have to be taken to ensure safety and avoid costly breakdowns. One example of a frustrating problem can be recollected from the oxidation of nuts and bolts. Those in the primary circuit were made of mild steel, which was satisfactory in all the earlier gas-cooled Magnox reactors. However, developments of these systems increased the temperature of the circulating carbon dioxide above the 340°C of earlier reactors and the oxidation rate of the steel became too high. The immediate solution was to down-rate the gas-cooled reactors which were affected so that the gas temperature was lowered, resulting in their operation at power levels as much as ten per cent below their planned output.

A notable feature of nuclear metallurgy has been in the field of what is known as post-irradiation examination. Materials can be removed from the reactor, sectioned, metallurgically examined or tensile strength tested without ever handling the metals or tools. Specimens are treated in vacuum furnaces, welds are made and X-rayed and photographs are prepared, all totally by remote control. Automatic machines have been developed which can remove fuel elements from reactors at various stages, by remote control, so that they can be measured and any unfavourable trends detected at the earliest possible stage.

Nuclear safety policy in Britain is based on four principles of defence in depth. Firstly the stations are run under international regulations, stringently controlled by independent safety experts. Secondly the power stations are designed to the highest known standards and inspected rigorously at each stage of construction. Quality control of the materials at all stages of manufacture is continued until each component of the reactor goes into service. Thirdly there is a comprehensive range of protective instrumentation which is triplicated or quadruplicated wherever necessary, to ensure that each reactor is brought to a safe condition in the event of a fault. Even the most remote possibility of an accident must be foretold and a routine established to deal with it. Finally, accidents in other countries are studied, including the devastating one at Chernobyl which led to much-needed, healthy international cooperation.

24

THE FUTURE OF METALS

We are thinking of the year 2000, over fifty years since the first edition of this book appeared. Possibly man's metallurgical endeavours may be dedicated to bigger and worse ways of self-destruction. However, it is more likely that common sense will prevail, so we will look at the future with some optimism. In order to discuss what metals will be needed, it may help to remind ourselves of man's basic needs – to sustain life, to make life comfortable, to have a measure of privacy but to be able to communicate in many ways with other people, to work, to enjoy leisure, to travel, to create and see beautiful things.

Assuming that we have scope to fulfil these needs, it is likely that more metals will be required for vehicles on land, on sea and in the air; for an increasing variety of domestic appliances; for all the girders, pipes, tubes, cables, wires and sheets that are required to make homes and places where people work or otherwise congregate; for the vast amount of agricultural machinery required by developing countries; for untold quantities of the mesh and rods which strengthen reinforced concrete and for the beautiful metal shapes that make bridges.

We anticipate that in the closing years of the twentieth century the world's production of iron and aluminium will double and this increase will be due to the economic expansion of previously undeveloped countries. China is 'on the move'; already the roads around Beijing, Shanghai and Canton are jammed with trucks. At present, however, there are only 270 000 cars in the whole country but they plan to have four million by the year 2000. With these and other developments, China plans to increase her steel output from the present 40 million tonnes to about 125 million tonnes by the turn of the century. In the western world the production of iron and steel may decline, not because of a lowering of the standard of living but because of more efficiently designed equipment and improved properties of materials.

The trend to reduce weight has been continuing since the Industrial Revolution. In 1810, when boilers were made of cast iron, the ratio of weight to power in a locomotive was 1000 kg per horse power. By 1900 the ratio was 100. When electric locomotives were introduced the ratio became 25 kg per horse power in the 1950s, and by 1980 it was only 14. Of course at the time this weight reduction was more than counter-balanced by the introduction of ferrous metals into many other new products – ships, bridges, buildings, automobiles and domestic equip-ment – but it is not likely that many new immense metal-using industries will develop.

Another reason for the fall in tonnage arises from the development of higher quality steels. In 1975 about 5 per cent of the steel in an average car was high strength or stainless; by 1985 the proportion was 14 per cent and it is expected to rise to 20 per cent by A.D. 2000. Such changes are accompanied by a reduction in the tonnage of steel, because each kilogram of high strength steel replaces about 1·3 kg of ordinary steel.

Although a similar trend will affect the consumption of aluminium in the western world, our tonnage will nevertheless increase because aluminium will be used in greater amounts in competition with ferrous metals, for example in automobiles. Even so, every individual manufac-tured item is bound to be weight-reduced. Thus the weight of each aluminium can decreased by about 20 per cent from 1979 to 1984, though the total weight used by that industry increased massively because of the greater number of aluminium cans produced.

SOURCES OF METALS

Helped by the development of transport and earth-moving machines, remote parts of the world in central Africa, western Australia and Brazil are being opened up to win new and rich sources of metals. The techniques which are evident in motorway construction have shown how large masses of earth can be dug out of one place and moved elsewhere, while at the same time landscaping the surrounding environment. In New Zealand, where they remove iron ore sand from beaches of North Island and turn it into a transportable slurry, the areas which have been depleted are afterwards re-landscaped.

Mining engineers estimate their resources of known ore deposits from the amounts which will be profitable to extract at current selling prices. Many forecasters allege that mining companies are only interested in proving that sufficient reserves exist for a mine to continue operating for thirty years and that there is no point in confirming the viability of reserves for a longer time. Consequently, economists and government officials looking into future supplies are often sceptical about the statements made by the so-called prophets of doom.

Much has been written about the forthcoming world shortages of metals. NATO and other organizations have identified metals, including copper, lead, zinc, silver, cadmium, cobalt, antimony, mercury and tungsten, which are likely to fall short of world requirements. They will become increasingly expensive and their use may be limited to specific applications for which they cannot be replaced.

Against this gloomy outlook for some metals we must add that, fortunately, deposits of iron, aluminium and magnesium ores will last for a very long time, with ample magnesium salts in the sea as an added bonus. Furthermore the ocean-bed is an exciting possibility as a supply of metal. Mineral nodules, like big pebbles, containing 5 to 30 per cent manganese, have been found on the ocean-bed in concentrations of 15 000 tonnes per square kilometre, at depths of over 3000 metres. The origin of these nodules is obscure, but deep-sea observation by TV cameras in the Galapagos area revealed two 'tectonic plates' of the earth's mantle drifting slowly apart. Lava from the interior is oozing out at very high pressure into the sea, which leaches manganese, iron and other elements from the lava; it is thought that these may then become ocean-bed nodules.

Preparations are being made to dredge areas between southern California, Nicaragua, Panama and Hawaii, the object being to obtain, not the manganese, but nickel, copper and cobalt. But does the ocean-bed belong to any one country or is it a common heritage of all mankind? Whoever claims that they own it will be faced with difficult and costly dredging and reclaiming operations.

In the first edition of this book there was a chapter discussing strategic metals in war-time. *Fig. 54* was drawn to illustrate the tug-of-war between the Allies, Germany, Italy and Japan. Even if nowadays military conflicts do not occur on a global scale, local wars seem to be inescapable and the contestants try to prevent each other from obtaining required materials. The effect of impoverished, militant or troubled countries on strategic supplies of metals can be confirmed by listening to the international news bulletins and then referring to the charts on pages 11 to 13, where we show the countries with the greatest output of metallic ores. For example, it will be seen that chromium is obtained from South Africa, Albania and Zimbabwe, while major supplies of vanadium ores come from South Africa and Namibia. Often the total economy of an otherwise under-developed country depends on its ability to excavate,

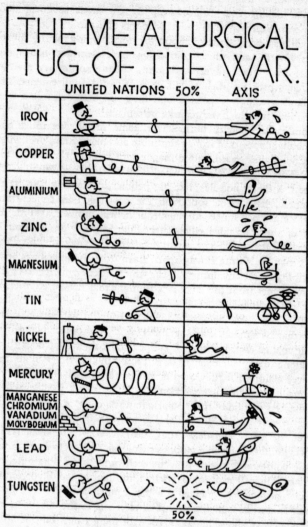

Fig. 54. The metallurgical tug-of-the-war, autumn, 1943 (*Reproduced from the first war-time edition of this book*)

transport and sell its ore supplies of a metal which is in demand by the rest of the world. The catastrophes endured by Zaire and Zambia in the late 1970s showed how such economies can be temporarily wrecked, resulting in worldwide shortages and enormous increases in the price of cobalt.

RECYCLING OF METALS

Before the turn of the century a number of new factors will determine the use of metals, especially those which are in limited supply. The collection, sorting and conversion of disused metal articles into re-processed ingots of steel, cast iron, aluminium and copper alloys is already a vast industry. Previously, the source of such material was from the rag and bone man's collection of old saucepans, bedsteads, broken-down lawn mowers and cheap tin trays.

A great deal of scrap comes from broken-down cars and domestic appliances; it has been estimated that each year, world-wide, about 20 million tonnes are obtained from these sources. After cylinder blocks and other heavy units have been removed, automobile scrap is fragmentized and this is followed by the 'Sink and Float' process, which separates and reclaims useful materials of different densities. The ferrous metals are first removed magnetically and the remainder is flooded with water, to take away pieces of rubber, dirt and other unwanted substances. Next, the material passes to tanks containing suspensions of magnetite and ferro-silicon in water, which can be adjusted in composition to specific gravities between 1·25 and 3·8. Flock and upholstery float on the lighter liquids, while the metallic substances sink. In a further stage, with fluids of the higher density, aluminium floats while zinc, brass and copper sink and are then separated manually.

Although the Sink and Float process provides a useful separation between the heavy and light metals, there is still the problem of the floating aluminium becoming entangled with other, mainly non-metallic, scrap such as ceramics, small stones, and bits of rubber, which have specific gravities similar to that of aluminium. This limitation has been overcome by a British invention which won the Prince of Wales Award for industrial innovation and production in 1986. In the Cotswold Separator, the aluminium and other fragments pass on a conveyor belt above a linear induction motor which forms an electromagnetic field at right angles to the line of the conveyor. The aluminium fragments are ejected away from the conveyor and are collected, while the unwanted non-metallic material continues and is finally dumped. In this way the aluminium contains only a small percentage of non-metallics; the process has the further advantage that very fine pieces that would not be collected in other processes are added to the total. The aluminium is converted to alloy ingots by secondary metal manufacturers and, since chemical analyses can now be undertaken within a few seconds by direct reading optical emission spectrometers, the composition of each melt can be adjusted to provide a desired specification.

Scrap of any metal in any country ought to be preserved and recycled with the same attention and control as if it were the metal in virgin condition. Yet often this is neglected. Taking a general view, humanity does a poor job of collecting and recycling its used metals and other materials. Several estimates in various countries have shown that only 15 to 25 per cent of each year's current production of the main tonnage metals is from old scrap. It should be above 50 per cent.

The increasing costs of accurately sorting metal scrap and converting it to good quality alloys will lead engineers and metallurgists to discuss the widening of specifications. There are cases where the accuracy of composition of an alloy must be controlled to fine limits, but in many other alloys too close a specification is pedantic. Alloys of aluminium and of copper, for example, can tolerate certain impurities in relatively large amounts without undue change in mechanical or other properties.

SOURCES OF ENERGY

About half the world's energy is used in the mining, extraction and fabrication of metals and other materials. Increasing consumption will require greater amounts of energy at a time when sources are becoming depleted. The amount of energy needed to make a metal is more than that used in the final smelting process. For example bauxite ores are quarried in tropical regions, transported, refined to aluminium oxide; cryolite is manufactured and the carbon anodes made; then about 14 000 kWh are required to electrolyse the oxide to a tonne of metal.

These processes use energy in the form of compressed air, oil and electricity and to make valid comparisons they must be converted to one unit, such as kilowatt-hours – a calculation that still causes controversy. That is not the end of the story, because when solid fuels or oil are burnt to make electricity in power stations, the process is only about 30 per cent efficient. Thus to obtain the 14 000 kWh for which the smelter pays, the total energy taken from the fuel amounts to 50 000 kWh. The other processes are added to that, making a total of over 70 000 kWh of energy per tonne of aluminium. Magnesium would require 115 000 kWh per tonne; in contrast, reinforced concrete requires 2300 and timber, which obtains energy free of charge from the sun, needs only 500 kWh per tonne.

There is another factor which must be taken into account: as rich and accessible ore bodies become depleted, more and more energy will be needed to win metals from the remaining deposits. For example, the metal content of available copper ores has decreased from about 6 per cent in the nineteenth century to 0·7 per cent now, and it is likely to fall as low as 0·2 per cent metal within the next few decades; thus the energy requirements per kilogram of copper will increase accordingly.

Nature provides some sources of energy free of charge – though the cost of collecting that energy is rather high at present. The most promising method of using solar electrical power is with photo-voltaic cells, consisting of semi-conductor diodes. The principle has been understood for the past thirty years but the real problem lies in making the apparatus cheaply enough. An operating cell is expected to give about 150 W of power per square metre; it is evident that we must be able to produce large areas of cells very cheaply for these to be competitive. The cost has been dropping quite steeply in the past few years but it is still greater than the optimum needed to compete with power station electricity or with oil. The fall in oil prices during the late 1980s made competition more difficult. In the discussions on aluminium we noted the advantages of hydroelectric power and we doubt whether solar energy will compete with a good dam for many years to come.

Feasibility studies are also being conducted to assess the possibility of extracting energy from tides and waves. When one or more methods have been selected, tested and put into operation, the size of the structures will be enormous, their possibilities immense and the use of metals comparable with the requirements of the shipbuilding industry.

The oil crisis of 1974 led to immense increases in the cost of energy and some companies which did not react quickly to the new problems went bankrupt. Others survived and began to investigate possible ways of preventing the waste of such a valuable commodity. Often their efforts began by calculating how much energy is required, theoretically, to operate each metallurgical process. For example, knowing the specific heat and latent heat of aluminium it is easy to calculate how many heat units – calories, therms or BTUs – are required to bring one tonne of aluminium to the temperature at which it can be cast. One tonne would require 353 million calories, or about half a million BTUs, which would be provided by 27 litres of fuel oil, 10 therms of gas or 294 kilowatt-hours of electricity. One can imagine the shock experienced on discovering that two or three times the theoretical amount of energy was being wastefully consumed.

Many solutions were found, including waste heat recuperation to superheat incoming air at oil burners, better insulation of furnaces, automatic furnace controls to cut down the heat once the required temperature had been reached and re-scheduling of processes to use low-cost electricity during the night. Apart from such improvements many companies realized that energy was being wasted by sheer foolishness: one consultant found that a 20 tonne charge of molten copper was held for several hours while an overhead crane was being replaced, whereas a re-scheduling of the repair would have prevented that waste of energy. Many discoveries of waste, such as leaking air compressors, were found to have caused surprisingly high costs.

NEW ALLOYS

In past years metallurgists tended to concentrate their researches on the major industrial metals: aluminium, copper, iron, magnesium, lead, nickel, tin, titanium and zinc, but increased prosperity and wonderful developments will come from the study of other metals. In the Dramatis Personae on pages xi–xii there were 36 metals and one non-metal (boron), followed by 23 rare metals, 12 very rare or synthetic metals, 4 radioactive elements, 3 'disintegration products' and 7 semi-metals and semi-conductors. Assuming that 67 elements in the list might be used in alloy systems, there would be over 4000 pairs to form binary alloys; perhaps 3000 of those pairs would have possibilities. If ternary alloys were considered, there would be some 180 000 combinations, while there would be over ten million possible quaternary alloy systems. The list would be enlarged by combinations with some non-metals such as carbon, phosphorus, hydrogen and nitrogen. In each alloy system there would be a range of composition from nearly zero to nearly 100 per cent, so to investigate the near-infinite number of possible alloys, metallurgical research should have a full programme ahead. Below, we mention just two developments in alloy metallurgy that will become increasingly important in the future.

In the small but vital field of magnetic materials several new alloy compositions are being developed, including some of the rare metals such as samarium–cobalt and boron–iron–neodymium. These can develop magnetic intensities more powerful than any previously known

Table 21. Probable future temperature ranges of materials

Material	Temperature °C
Tungsten-base and tantalum-base alloys	1500–2000
Molybdenum-base alloys	1200–1400
Niobium-base alloys	1100–1350
Chromium-base alloys	to 1200 but limited to very low stresses above 1000
Complex alloys containing two or more of the above metals as a base	1650
Nickel-base and cobalt-base alloys	1000
Iron-base alloys	900
Titanium-base alloys	600

alloys and are leading to dramatic improvements in the efficiency of electric motors and videotape recorders.

Lithium is the lightest of all metals but it is highly reactive. Although experimental work on aluminium–lithium alloys was begun in the 1950s, early work was not successful. Recent developments have overcome the production problems and bulk output is being achieved, especially in an aluminium alloy containing 2·5 per cent lithium, 2·4 per cent copper and 0·75 per cent magnesium, with 0·12 per cent zirconium for grain-size control. Such an alloy is ten per cent lighter and ten per cent stiffer than the conventional high-strength aluminium alloys, thus enabling the designer to make up to 15 per cent weight-saving in a structural component or assembly. At present the aluminium–lithium alloys are several times more costly than the competing aluminium alloys but their advantages justify their effective use in space satellites and for the airframes of supersonic aircraft.

ALLOYS FOR HIGH-TEMPERATURE SERVICE

Molybdenum, tungsten, tantalum and niobium alloys will be used in appropriate temperature ranges, sometimes alone, sometimes coated with a thin layer of ceramic. Some estimates of the temperature ranges at which various alloys can be used are given in *Table 21.*

The properties of materials with high strength at elevated temperatures can be utilized efficiently by making honeycomb structures with thin strips and progressively welding or high-temperature brazing them, so that abutting walls are joined into a honeycomb form. This procedure is already exploited for 'skinning' the fuselages of aircraft and space vehicles, because of the great strength and rigidity obtainable, coupled with a high capacity to absorb vibration.

Already the demand for materials capable of withstanding high temperatures has resulted in the development of a new family of materials which contain a metal combined with carbon, nitrogen, silicon or boron. Small particles or fibres of these refractory compounds are held together with a ductile metal or a ceramic acting as a cement. Being a combination of ceramic and metal, these materials are known as 'cermets' and are formed into blocks, closely approaching the shape required for their end-use.

METAL MATRIX COMPOSITES

When a material is formed into a wire or fibre its strength per unit cross-section is much greater than when it is in bulk. Research laboratories in many countries are producing metal matrix composites, in which a strong alloy, usually of aluminium or magnesium, is made immensely stronger by fibre reinforcement. In Britain, large aluminium alloy castings reinforced by as much as 65 per cent by volume of high-strength boron fibre are being made for aerospace requirements. The fibres are usually produced by vapour deposition; for example, boron is deposited on a tungsten filament, followed by a surface coating of silicon carbide or boron carbide to increase the bond strength between fibre and matrix and to prevent oxidation.

There is a great variety of fibre materials available such as graphite, silicon, boron, alumina and silicon carbide. The composite materials are expensive but, as the launch cost of satellites ranges from 2000 to 20 000 dollars per kilogram, a small saving in weight more than justifies the cost of using composites, which are strong and light, thus making weight reductions possible. One satellite has over four thousand composite components. Their use is extending to ground-based industries, such as the Japanese mass-production of diesel engine pistons.

COMPETITORS OF METALS

Why does man use metals – or timber or concrete? Apart from aesthetic considerations a material is used because it offers the required strength and other properties at minimum cost. In working out the strength factors we need to know the cost per kilogram, the density of the material, and its strength per unit of cross-section. When manufacturers obtain a given weight of metal, they subconsciously visualize its final form as a girder or the connecting rod of an automobile, for example. Thus they are purchasing with cost per unit strength instinctively in mind. In other situations, say with overhead electrical cables, they assess the cost per unit of electrical conductivity coupled with strength.

For most large engineering structures strength is the most important consideration. It is not surprising, therefore, that because reinforced concrete is the cheapest way of buying strength, it is far and away the major tonnage and volume material used by mankind. Of course, reinforced concrete depends on its strengthening, hidden core of steel. Second is timber, which is a remarkably cheap way of buying strength. All over the world vast areas are being planted, or replanted with trees because the sun's rays provide the energy of 'manufacture' of timber and it is a material that can give great strength and a very long life. The third best bargain is structural steel, followed by aluminium and plastics, in that order. In the future the energy required to manufacture a material, compared with the strength or electrical conductivity, which that material provides, will be a paramount consideration.

Metals and plastics can often be combined. For corrosion-resistance, steel pipes lined with plastic are being used on an increasing scale, and plastic-coated steel is used for building. Structural partitioning is made by bonding wood dust with plastic, and coating it on both sides with aluminium strip. But the combination of metals with plastic brings a problem of reclamation even more difficult than the recycling of plastics.

REFRACTORIES AND CERAMICS

Since the earliest days of metallurgy, refractories have played a vital part in metallurgical operations, because without them it would have been impossible to use the high temperatures required for the efficient extraction, refining and casting of metals. Whereas, in the past, refractories have been considered as ancillary to metal production, they are now taking on a new interest and becoming important 'end products' combined with metals. Requirements of jet turbine blades to operate at high temperature are so exacting that, in the not too distant future, a mixture of refractory metal oxide bonded with a metal may exhibit better service behaviour than a heat-resistant alloy.

The ceramics industry has always been aligned closely to metallurgy; their problems are somewhat akin and the one is dependent upon the other. In recent years, a closer tie between ceramics and metals has developed because of the need for materials which are capable of carrying stresses at much higher temperatures than has been hitherto necessary. Such materials are now required for heat engines, space missiles and nuclear reactors.

LASERS

Young metallurgists would be well advised to learn about the construction, operation and possibilities of lasers for manufacture and research. A few notes on the different types of lasers appear on page 295. Low power lasers are used for communications; those with up to 100 W are employed for drilling holes, scribing and engraving metal articles. Lasers of from 100 to 1500 W power are suitable for cutting the panels of cars and aircraft, for heat treatment of such parts as camshafts and for welding, as was discussed on page 244. In metallurgy, laser surface alloying is an active topic and there have been advances in the use of lasers for metal deposition, machining and the measurement of residual stresses, for example in electro-deposited coatings.

Lasers can cause the separation of isotopes; a beam is tuned to the frequency of the atom (or more often the molecule containing the atom) of one of the isotopes of an element. This enables us to separate the material containing one isotope from that of another, which would require a different 'tuning'. Perhaps this will add to the dangers of life by making it easier to separate the uranium 235 isotope from the 99·3 per cent of the remainder, but it is to be hoped that other more valuable achievements will come from this development.

25

METALLURGY AS A CAREER

A generation ago metallurgists were more or less confined to their laboratories, so opportunities for advancement were not plentiful. The general public had little understanding of the functions of a metallurgist, indeed many people thought that he was connected with weather forecasts. Only a few years ago, a class of schoolchildren was asked to define a metallurgist; most of them did not know the meaning of the word, though one hazarded a guess that he was 'a small animal, nice and furry'. The recognition of the importance of metallurgy has come gradually and is still continuing, but now metallurgy as a career opens up prospects of management and of being in charge of processes, instead of being an advisory boffin. The opportunities in the future will lie in several fields.

MINING AND PRODUCTION

All over the world labour costs are increasing and the cost of energy is becoming greater. In the 1970s there was a belated realization that existing cheap sources of energy were limited and that new sources will take years to develop. Many prices of materials increased substantially and some metals became scarce. These higher prices led to the exploitation of new deposits, a reappraisal of weak ores and the development of extraction processes suitable for those ores. At the same time the increased value of by-products had a useful effect on the profitability of some smelting processes; for example small percentages of silver and mercury, obtained during zinc smelting operations, helped these processes that were not very profitable to become more viable. The Bingham Canyon copper mine, which was referred to on page 15, produces molybdenum and gold as by-products. It is expected to become one of the world's largest gold producers by 1992.

While these developments have been affecting the economics of metal smelting, problems of contaminating the environment have become of major concern. Fume emissions, working conditions and good relationships with local inhabitants will henceforth affect the future of any new or existing metal smelter. This can only be done at great expense, which may lead to uneconomic manufacture. The rise in zinc prices in the mid-1970s was caused partly by the closing of five major American units, because of contamination of the environment. The efforts to produce metal economically have been linked with new methods of discovering ore deposits, excavation, transport and concentration of ores, and new methods of smelting the ores with less energy per tonne than before. Such endeavours, coupled with a more efficient recycling of scrap metals, will become one of the biggest challenges for the metallurgist of the future.

THE METALLURGIST IN THE CONSUMER INDUSTRIES

In this field, firms which design, engineer or construct will employ the metallurgist or materials technologist (it is a pity that, so far, nobody has thought of an acceptable single word for this). These 'materialogists' will be specialists in the melting, alloying, shaping, treatment and service behaviour of ferrous and non-ferrous metals. They will regard other materials – concrete, rubber, plastics, timber – as requiring the same kind of study as metals. Part of their training will be in Value Analysis in which the choice of materials, their dimensions and assembly, are scrutinized from first principles.

The nuclear power metallurgist can be included in this group, though the metals that once were classified as exotic are widely used in nuclear engineering and need a study of their own. Furthermore the paramount needs of safety bring problems that were never encountered before. When eventually atomic fusion, as distinct from fission, is harnessed for creating power, a new series of metallurgical problems will arise.

There are many opportunities for the metallurgist in the fields of high technology. Materials with the greatest possible strength, combined with resistance to heat and minimum weight are required for aircraft and all kinds of space vehicles. This involves developing the use of new alloys and composite materials, offering exciting challenges to the metallurgist.

TRAINING FOR MANAGEMENT

The opportunities for metallurgists in management are developing rapidly. This is a great change for the better; not so long ago, the metallurgist, though qualified and necessary, was out on a limb and did not get many opportunities to progress directly into management. This was partly caused by the training of metallurgists a generation ago. The authors of this book remember being sent, when they were students, for a fortnight each year to whichever metal works would condescend to accept them as visitors. This training usually developed into helping with simple analyses in the laboratory, helping to pay the wages on Fridays or fishing in the nearby canal because no suitable work could be found.

University courses are now designed to make students aware of the opportunities and responsibilities in management. It is not uncommon for undergraduates to work for several months with metallurgical companies. Their duties are discussed before and during the course with the University staff, the employers and the student. On return he or she will have gained a better idea of 'what it is all in aid of' and will have observed some of the problems which will be encountered later.

The study of management, including cost accountancy, is entering more and more into University studies. Students are now required to design plant layouts for a given process, to estimate the cost of the building, machinery, power and materials and to link their studies of the project with visits to factories which are concerned with the same problems.

There will always be scope for the manager-metallurgist in large firms and increasing opportunities are arising where the combination of technical knowledge, cost consciousness and management training can equip a young person for promotion in a company. Great opportunities exist for women in metallurgy. No doubt a generation ago the heat of processes, the danger of splashing metal and the robustness of the language combined to make women feel out of place in the 'man's world' of metallurgy. However, processes are becoming automated and computerized; mathematics, electrical engineering, electronics and other disciplines are associated with metallurgical processes. Conditions of work and opportunities are now very acceptable to women. When the authors were students in the 1930s, just one shy young lady became a metallurgical student and she remained for only two years. Nowadays women account for 25 per cent of the intake in metallurgical departments in universities. Women are recognized as highly motivated and they achieve excellent results in the final examinations; there are many examples of very successful subsequent careers.

Managers of the future will be involved more and more in problems of ecology and safe working conditions. One prevalent cause of accidents in the metallurgical industries has been – and still is – the movement of metal, whether molten or solid. In spite of control devices on fork-lift trucks and automatically operated ladles, accidents are still happening. An ever-present worry for those who are training young workers is the explosion that occurs if molten metal is poured into a mould that is moist. Only the most rigorous instruction and insistence that safety glasses and protective clothing are worn will eliminate such dangers. Safety starts with the engineers who design guarding and similar equipment and finishes with the Safety Committee which makes sure that the equipment is used to the full; but it is essential that managers give full backing to all these endeavours.

For many years the metal industries were noisy, particularly in fettling shops where grinding machines, mechanical chisels and shot-blasters made it almost impossible to converse without lip-reading. All this has been changed as a result of noise abatement legislation, coupled with the requirement that operators who work in noisy conditions must have hearing protection. Local communities near factories are conscious of their right to demand noise reduction, particularly at night, and to insist that toxic and unpleasant fumes be prevented from spreading into the atmosphere.

THE METALLURGIST AS SHERLOCK HOLMES

The examination of failures or flaws is an interesting and challenging service. Investigation of motor-car accidents, exploded steam boilers or bridges that have collapsed involve retesting and microstructural examination of the parts and much work of a forensic nature. An important spin-off is that examinations of failed structures increase our total experience and knowledge, leading to improved design, manufacture and construction and hence prevention of failure. If the destructive effect of harmonic oscillations of the Tacoma Bridge, the potential weakness of box girder design and the stresses undergone by the DC 10 engine-mounting assembly (which may have led to the Chicago air disaster in May 1979) had been foreseen, much suffering, loss of life and waste of money would have been avoided.

Metal fatigue is an ever-present hazard, especially when a structure operates in corrosive conditions. The Norwegian oil platform *Alexander Kiellan* suffered a catastrophe which led to the loss of over a hundred lives, because what was thought to be a minor modification was undertaken without considering its possible effect. The platform was supported on vertical legs, held by steel tube braces about a metre in diameter. It was decided to drill a hole in one of the braces, so that some communication equipment could be fitted; after that the brace was welded. It developed a crack which grew to over a metre long, causing the collapse of the brace and then the whole structure.

METALLURGICAL RESEARCH

When the authors were beginning metallurgical research it was necessary to spend many weeks reading scientific journals, hoping to discover some information which might be useful. Nowadays the computerized library provides an immediate, up-to-date and concise report on any aspect of research from countries all over the world.

Scientists are provided with measuring instruments of almost unbelievable accuracy. It is possible to measure time intervals lasting less than one millionth of a millionth of a second, the time it takes for a ray of light to travel about one millimetre. The capability of X-ray examination has been extended by neutron radiography; it is especially suitable for examining liquids and hollow structures such as the rotor blades of gas turbines. The electron microscope has made it possible to isolate and examine the behaviour of minute groups of atoms and molecules. American physicists have made 'quantum layers' and 'quantum dots' of gallium arsenide containing alternating layers of indium and aluminium atoms. These structures, about 750 atoms across, measure only a hundred millionths of a millimetre. One laboratory achieved the isolation of ten atoms. Such minute groups are found to have properties quite different from larger accumulations of the same material; for example they become superconductive. These developments first opened up new fields in semi-conductor technology, but now they are becoming significant in metallurgical research.

In March 1987, three thousand physicists crammed into a 1200-seat auditorium in New York; the meeting began at 7 pm and finished at dawn and during it scientists described their researches into superconductivity. Many talks were given at that meeting, including one by the Swiss researchers who discovered the new superconductive materials and who later gained a Nobel Prize. The first information had been published a year earlier, but was very little noticed until the results were confirmed by a Japanese group towards the end of 1986. American scientists made more developments and by March 1987 many people were working on the new discoveries.

Superconductivity, which allows electricity to flow without any resistance, has been known since the 1930s but the phenomenon occurred only at temperatures a few degrees above absolute zero, requiring expensive liquid helium to reach that temperature. Consequently the then-known superconductive materials, including niobium–tin mentioned on page 214, were used only for sophisticated equipment such as very powerful electromagnets where the high cost was justified.

The newly discovered materials are known as perovskites – they are ceramics containing copper, with each copper atom surrounded by six oxygen atoms, arranged in layers, with either barium and lanthanum or barium and yttrium in between each layer. The perovskites become superconductive at minus 150°C, a low temperature that can be reached using the comparatively cheap liquid nitrogen. Since the original discoveries an ever-increasing number of materials have been tested in the hope of bringing the operating temperature of superconductors a little nearer to room temperature. For example the Bell laboratories in New Jersey have experimented with superconductors based on lead, strontium, lanthanum, calcium and copper oxide. They also developed another group with barium, potassium, bismuth and oxygen; this was the first not to be based on copper oxide.

Almost every month sees new discoveries in this field, opening the way for revolutions in computers, nuclear fusion, medical scanners and the new 'superconducting collider' which the U.S. is planning to build at a cost of five billion dollars. Superconductors repel magnetic fields, so they will stay suspended above a magnet and thus will be used to power high-speed levitating trains which run on magnetic fields instead of rails.

Another development that will affect metallurgical processes in the future is known as rapid solidification technology (RST). The key to RST is the extremely fast cooling of an alloy from liquid to a powder. The rate of cooling is of the order of one hundred degrees per one ten-thousandth of a second. In one method a fine spray of metal droplets falls on to a rapidly spinning wheel, which throws them out into an extremely cold atmosphere. In another method a laser beam is passed rapidly over the metal's surface, forming thin layers of molten material which immediately become frozen by the bulk of the solid. Applications of this 'laser glazing' include hardening the tips of turbine blades.

Research on metallurgical processes is continuous, always stimulating, though often disappointing when a new process that went well in the laboratory develops problems in full-scale production. The direct reduction of iron ores to steel, continuous processes for making sheet metal, new smelting processes and the extension of powder metallurgy to the making of large components are a few of the multitude of projects.

The state of mind needed in industrial research is what we might describe as 'the man from Mars' approach. Such a man might visit a manufacturing company and, because he had never seen the process before and had no preconceived ideas, he might notice the illogicalities or follies of that process. Those who are concerned every day with a process are often limited by immersion in their local problems.

In research on processes there are at least three stages, each of which needs this 'man from Mars' approach coupled, ideally, with a down to earth attitude. First the process is worked out on a small scale under laboratory conditions to establish the principles. Next a medium-sized prototype plant is built, operating perhaps in the works but still under careful control and measurement. Finally the large capital expenditure for a full scale plant is authorized. As the size of the process increases, new problems arise, which need an objective and self-critical attitude. For example, doubling the linear dimensions of a container multiplies its volume by a factor of eight or, as we remarked when discussing Japanese blast furnaces, doubling the inner surface area quadruples the volume. Sometimes increase of scale is beneficial, as with iron blast furnaces; sometimes it brings environmental problems in its train, as happened with the blast furnaces in the new smelting process for zinc. In all developments changes in size or method require clear and unprejudiced thinking. There are great opportunities for young and clever scientists, metallurgists and engineers who realize that, apart from devising a new process, the perfecting of it also requires unmitigated attention.

GLOSSARY

(Technical terms which are indexed and amply defined in the text are not included in the glossary.)

AGEING. As applied to castings in steel and cast iron, the word indicates a period of time provided to relieve casting stresses. Ageing is also used in reference to wrought aluminium alloys of special composition; after heat-treatment they are quenched in water and then kept at room temperature for a period of four to six days. During this period, their maximum strength and mechanical properties are fully developed.

ALPHA PARTICLE. The nucleus of the helium atom, containing two neutrons and two protons, emitted from the nuclei of certain radioactive elements.

ÅNGSTRÖM UNIT. A unit of measurement used by metallurgists and crystallographers, giving the distance between atoms in a space lattice. An Ångström unit is one ten-millionth of a millimetre.

ANNEALING. The process of heating metal or alloy to some predetermined temperature below its melting point, maintaining that temperature for a time, and then cooling slowly. Annealing generally confers softness.

BINARY ALLOY. An alloy composed of only two major ingredients, e.g. copper and zinc, or lead and antimony. An alloy with three ingredients is known as a 'ternary alloy'.

BREAKING DOWN. The first stage in the shaping of an ingot of metal, with the object of reducing its section and refining its grain structure.

CARBIDE. The compound formed when an element combines with carbon. The carbides of metals are usually intensely hard.

CATALYST. A substance which, when present in small amounts during a chemical reaction, promotes the reaction, but itself remains unchanged.

CEMENTITE. The name given to identify one constituent in iron–carbon alloys.

Cementite is essentially iron carbide, Fe_3C, but may contain other elements such as manganese and chromium, carbides of which are dissolved in the iron carbide and do not appear separately.

CERMETS. Materials produced by bonding a metal oxide, carbide, nitride, or boride, with a ceramic. The bonding is effected at high temperature under controlled conditions, by methods similar to those used in powder metallurgy.

COMPATIBILITY. A term used especially in nuclear engineering. Two materials are said to be compatible when they can exist in contact with each other without interaction.

CONDUCTIVITY (ELECTRICAL or THERMAL). The measure of the ability of a substance to allow the passage of electricity or heat. Copper is an example of a good conductor, rubber of a bad one.

CORES. Specially fashioned pieces of sand or metal used to form the hollow parts of a casting. To make a cylindrical hole in a casting, a cylindrical solid core of similar shape is used.

CRYOGENIC. The condition of materials and the behaviour of phenomena at very low temperatures.

DEEP DRAWING. A shaping process in which the whole or part of a disc of metal is forced through the aperture of a die so as to make a cup shape. By repeating this process with plungers and apertures of progressively decreasing diameters, the metal is drawn into an elongated cup or closed tube.

DIELECTRIC. A non-conductor of electricity.

DIES. Metallic or other permanent forms which confer a given shape on a piece of metal. The word covers a range of meaning and includes dies which are used to shape solid metal in presses and those for making diecastings from liquid metal.

DISTILLATION. The conversion of the whole or part of a liquid substance into gaseous form, to be followed by the subsequent condensation of this to liquid.

DOPING. The addition of a small amount of impurity to a semi-conductor material, to make an immense change in its electrical conductivity. A typical dopant is boron added to silicon, in the proportion of between one and a hundred parts to a million.

DUCTILITY. The property of a metal which enables it to be given a considerable amount of mechanical deformation (especially stretching) without cracking.

ELECTRODE. A conductor which conveys electric current directly into the body of an electric furnace, plating vat or other electrical apparatus.

ELECTROLYSIS. A process involving chemical change, caused by the passage of an electric current through a fluid solution.

ELECTRON. Elementary negatively charged particle having a mass about 1/1840 that of a hydrogen atom.

ELECTRON MICROSCOPE. This is a form of microscope which is capable of very high resolution. For the detection of the finest details of a structure, light, the exploring medium of an optical microscope, has too coarse a 'texture' (wavelength). The electron microscope employs an exploring medium of a much finer 'texture' – a stream of electrons. The stream can be rendered parallel or convergent by 'lenses' consisting of magnetic fields. The electron stream is scattered by the object to be examined and an image is formed on a fluorescent screen or on a photographic plate. Resolutions of over ten million can be obtained.

ELEMENTS. All compounds can be resolved into elements; thus, water into hydrogen and oxygen; common salt into sodium and chlorine. The elements, however, cannot be resolved by chemical means into any simpler substances. Including the modern 'synthetic elements' which have been made in the atomic physicist's laboratory there are over a hundred elements, of which more than three quarters are metals.

EQUILIBRIUM. The state of balance which exists or which tends to be attained after a chemical or physical change has taken place. Equilibrium may not be reached for long periods after the change has been initiated.

FLASH. When a metal is forged or cast in a die some metal penetrates the space where two die surfaces touch and thus a web of metal, known as 'flash', remains attached to the forging or casting and has to be removed by filing or clipping.

FLUX. A chemical, used to combine with a substance having a high melting point, generally an oxide, forming a new compound which can readily be melted.

GAMMA RADIATION. Rays of very short wavelength, less than that of X-rays, produced during the disintegration of radioactive materials.

HOT SHORTNESS. An undesirable property of certain metals and alloys whereby they are brittle in some elevated temperature range.

HYPERBARIC WELDING. A method of producing welds underwater in a dry habitat at greater than atmospheric pressure. The further below the surface the position where the weld has to be made, the higher the pressure in the working chamber, since the pressures must be sufficient to displace water within the chamber.

INCLUSION. A non-metallic particle of slag, oxide, or other chemical compound which has become entangled in metal during its manufacture.

INDUCTION FURNACES. The metal is held in a refractory container surrounded by a coil through which alternating current passes. This induces currents in the metal, causing it to be heated and, if required, to melt by internal resistance.

The frequencies of the applied currents vary, depending on the amount of metal. Small quantities in laboratories are melted in high frequency furnaces ranging from 10 000 to a million cycles per second. Medium weights of metal, for example half a tonne of aluminium alloy melted in a foundry, require a frequency of 1000 cycles. Very large tonnages are melted in low frequency furnaces operating at 50 cycles per second – the same as that used in Britain for domestic supply.

INGOTS. Blocks of metal made by casting the liquid metallic content of a furnace or crucible into open metallic moulds.

ION. An electrically charged atom or group of atoms.

ISOTOPES. Atoms of the same element, having the same number of electrons and protons but different numbers of neutrons. The isotopes of an element are identical in their chemical and physical properties except those determined by the mass of their atoms.

LASER (Light Amplification by Stimulated Emission of Radiation). A laser consists of three components. The material which will be made to 'lase' can be either a gas (carbon dioxide produces the most powerful beam), a liquid (such as a dye dissolved in ethylene glycol) or a solid (ruby was the first laser material but gallium arsenide will probably be most widely used in the future). The second component is a source of excitation, which can be electric power or light. The third component is an amplifying arrangement which normally consists of two parallel mirrors positioned so that the light can bounce backwards and forwards between them, becoming amplified in strength as it passes each time through the laser material. The amplified laser beam produced is highly directional and usually of one wavelength.

LATERITIC ORES. The word is derived from the Latin for 'brick'. Red-coloured clay-like laterite materials are used for road building in the tropics. Lateritic ores, containing small amounts of nickel, occur in equatorial parts of the world. They contain much moisture and compounds of iron which have to be removed, with difficulty, before the ore is in a suitable condition for smelting.

MACH NUMBER. The relation between the speed of a moving body to the local speed of sound, which can vary with temperature, altitude, and therefore pressure. The terms are generally used in connection with the speeds of supersonic aircraft. Mach 1 is the speed of sound at sea level; the speed of sound is reduced with increasing height. The word is derived from the name of Ernest Mach, a nineteenth-century mathematician and philosopher.

MAGNETIC PERMEABILITY. The ratio of the strength of magnetism in a material to the strength of the external magnetic field which induces it.

MALLEABILITY. A property of metals enabling them to be hammered and beaten into forms such as that of thin sheets, without cracking. Gold is the most malleable of all metals.

MEGAWATT. One million watts; one thousand kilowatts.

METALLOGRAPHY. The study, observation, and photographing of the structure of prepared specimens of metals, usually with the aid of a microscope. From such a study, much can be learnt about the condition, heat-treatment, and manufacturing history of metals.

METALLOID or SEMI-METAL. An element which has some properties characteristic of metals, others of non-metals. Examples are arsenic and antimony.

METALLURGY. The art and science of producing metals and their alloying, fabricating, and heat-treatment. The word can be pronounced 'metall*u*rgy' or 'met*a*llurgy', though the former may be preferred in keeping with the adjective 'metall*u*rgical'.

MICRON. A millionth part of a metre: a thousandth part of a millimetre, 10 000 Ångström units.

MODULUS OF ELASTICITY. The ratio of stress to strain in a material. The strain is usually a measure of change of length. The stress is a measure of the force applied to cause the strain. (*See* Young's Modulus.)

NEUTRON. Electrically uncharged particle possessing a slightly greater mass than the proton. Neutrons are constituents of all atomic nuclei except that of the normal hydrogen atom.

OXIDE. A chemical compound formed when an element unites with oxygen, as by the action of burning. Water is hydrogen oxide; sand, silicon oxide; and quicklime, calcium oxide.

PIG IRON. Crude iron as produced from the blast furnace and containing carbon, silicon, and other impurities. On flowing from the furnace the molten iron is run into channels which branch out into a number of offshoots about $1300 \times 130 \times 130$ millimetres. The iron runs into these and takes the familiar form of 'pigs'. The main channels are called 'sows'.

PROTON. Positively-charged particle in the nucleus of an atom having a mass about 1840 times that of the electron and an electric charge equal in magnitude to the negative charge of the electron.

PYROMETERS. Instruments for measuring high temperatures.

REAGENT. A substance which is added to another in order to bring about and take part in chemical action.

REFRACTORIES. Firebricks or other heat-resisting materials used for lining furnaces and retaining the heat without allowing the outer shell of the furnace to be damaged. Refractories are grouped into 'acid', 'basic', and 'neutral' according to their composition and their action on the hot substances with which they come into contact in the furnace. Examples of the three types are silica, dolomite, and carborundum respectively.

RETORT. A specially shaped vessel intended for containing substances which are to be heated to form a vapour which is then collected or condensed.

ROASTING. The process of heating an ore at medium temperature in contact with air.

SEMI-CONDUCTORS. Materials which at room temperature have much lower electrical conductivities than metals, but whose conductivities increase substantially with increase of temperature. This is in contrast with the electrical behaviour of metals, whose conductivities decrease slightly with increase of temperature. Silicon, germanium and selenium are semi-conductors. They derive their importance from the fact that their conductivity is extremely sensitive to the presence of impurities. Almost all electronic devices in use at present include semi-conductors.

SLAGS. Glass-like compounds of comparatively low melting point, formed during smelting when earthy matter contained in an ore is acted on by a flux. If the earthy matter were not deliberately converted into slag it would clog the furnace with unmelted lumps. The fusibility and comparatively low density of the slag provides a means by which it may be separated from the liquid metal.

SMELTING. The operation by which a metallic ore is changed into metal by the use of heat and chemical energy.

SOLUTION. The intermingling of one substance with another in so intimate a manner that they are dispersed uniformly among each other, and cannot be separated by mechanical means. Although solid substances dissolved in water are the most familiar solutions, the word has a wider meaning; for example, gases dissolved in liquids, liquids dissolved in liquids, and, frequently in metallurgy, solids dissolved in solids.

SPECIFIC GRAVITY. The ratio of the density of a substance to the density of water.

SPECTRUM. The result obtained when radiation is resolved into its constituent wavelengths. The spectrum of white light, consisting of coloured bands, is the most familiar, but all types of radiation comprising a range of wavelengths can be resolved into spectra.

STRAIN. A measure of the amount of deformation produced in a substance when it is stressed.

STRESS. A measure of the intensity of load applied to a material. Stress is expressed as the load divided by the cross-sectional area over which it is applied.

TAPPING. The controlled removal of liquid metal or slag from a furnace, the metal running through a tap-hole near the base of the furnace. In engineering the word denotes cutting an internal screw-thread.

TECTONIC. An architectural term relating to the structure of buildings. In geology the word signifies the structure of the earth's crust.

TEMPERING. A warming process intended to alter the hardness of a metal which has already been subjected to heat-treatment. The tempering temperature is lower than that at which the first heat-treatment was carried out.

THERMIT. A mixture of powdered iron oxide and aluminium which, when ignited, sets up a vigorous chemical action, whereby the aluminium unites with oxygen from the iron oxide, leaving metallic iron. A high temperature develops, so great that the iron is melted.

THERMOCOUPLE. When two wires of different metals are joined at each end and one junction is heated, a small electric voltage is produced which depends on the temperature and which can be measured by a delicate instrument. Certain combinations of metals are specially suitable, by virtue of the amount of current produced and their resistance to heat. Thermocouples are used for some types of pyrometers.

THERMO-PLASTIC. A plastic material which can be melted or softened repeatedly without change of properties. Injection mouldings are made with thermo-plastic materials.

THERMO-SETTING. A plastic material which can be softened by the combination of heat and pressure and then be compression-moulded into a required shape. Once having been treated in this way the thermo-setting material does not revert to its former condition and cannot be re-treated.

TROY. System of weights used for gold and silver. A kilogram contains 32·15 troy ounces.

ULTRASONIC. Pressure waves of the same nature as sound waves but whose frequencies are above the audible limit.

X-RAYS. Radiation of a character similar to that of light, but of a shorter wavelength and possessing the property that it can pass through opaque bodies.

YOUNG'S MODULUS. The ratio of applied stress to the strain which occurs when a metal is subjected to tension or compression (also known as the modulus of elasticity). It was named from Thomas Young, F.R.S., a genius in medicine, physics, optics and Egyptology who died 160 years ago.

ZONE REFINING. A method of purifying crystalline materials. A bar of the solid substance is progressively moved through a furnace in such a way that there is a small molten zone; during subsequent solidification some of the impurities diffuse towards the portion last to solidify. By repeating this operation progressively the impurities become segregated at one end of the bar, which is then removed. Metals can be zone refined to contain less than one part in a million of impurity; germanium has been produced with only one part of impurity in a hundred thousand million. The technique can be used for non-metallic materials and is an essential part of the manufacture of the super-pure silicon used in electronic devices.

INDEX

FOR THE BEST IN PAPERBACKS, LOOK FOR THE

In every corner of the world, on every subject under the sun, Penguin represents quality and variety – the very best in publishing today.

For complete information about books available from Penguin – including Pelicans, Puffins, Peregrines and Penguin Classics – and how to order them, write to us at the appropriate address below. Please note that for copyright reasons the selection of books varies from country to country.

In the United Kingdom: Please write to *Dept E.P., Penguin Books Ltd, Harmondsworth, Middlesex, UB7 0DA*

If you have any difficulty in obtaining a title, please send your order with the correct money, plus ten per cent for postage and packaging, to *PO Box No 11, West Drayton, Middlesex*

In the United States: Please write to *Dept BA, Penguin, 299 Murray Hill Parkway, East Rutherford, New Jersey 07073*

In Canada: Please write to *Penguin Books Canada Ltd, 2801 John Street, Markham, Ontario L3R 1B4*

In Australia: Please write to the *Marketing Department, Penguin Books Australia Ltd, P.O. Box 257, Ringwood, Victoria 3134*

In New Zealand: Please write to the *Marketing Department, Penguin Books (NZ) Ltd, Private Bag, Takapuna, Auckland 9*

In India: Please write to *Penguin Overseas Ltd, 706 Eros Apartments, 56 Nehru Place, New Delhi, 110019*

In Holland: Please write to *Penguin Books Nederland B.V., Postbus 195, NL–1380AD Weesp, Netherlands*

In Germany: Please write to *Penguin Books Ltd, Friedrichstrasse 10–12, D–6000 Frankfurt Main 1, Federal Republic of Germany*

In Spain: Please write to *Longman Penguin España, Calle San Nicolas 15, E–28013 Madrid, Spain*

In France: Please write to *Penguin Books Ltd, 39 Rue de Montmorency, F-75003, Paris, France*

In Japan: Please write to *Longman Penguin Japan Co Ltd, Yamaguchi Building, 2–12–9 Kanda Jimbocho, Chiyoda-Ku, Tokyo 101, Japan*

Illusions Charlotte Vale Allen

Leigh and Daniel have been drawn together by their urgent needs, finding a brief respite from their pain in each other's arms. Then romantic love turns to savage obsession. 'She is a truly important writer' – Bette Davis

Snakes and Ladders Dirk Bogarde

The second volume of Dirk Bogarde's outstanding biography, *Snakes and Ladders* is rich in detail, incident and character by an actor whose many talents include a rare gift for writing. 'Vivid, acute, sensitive, intelligent and amusing' – *Sunday Express*

Wideacre Philippa Gregory

Beatrice Lacey is one of the most passionate and compelling heroines ever created. There burns in Beatrice one overwhelming obsession – to possess Wideacre, her family's ancestral home, and to achieve her aim she will risk everything: reputation, incest, even murder.

A Dark and Distant Shore Reay Tannahill

'An absorbing saga spanning a century of love affairs, hatred and high points of Victorian history' – *Daily Express* 'Enthralling . . . a marvellous blend of *Gone with the Wind* and *The Thorn Birds*. You will enjoy every page' – *Daily Mirror*

Runaway Lucy Irvine

Not a sequel, but the story of Lucy Irvine's life *before* she became a castaway. Witty, courageous and sensational, it is a story you won't forget. 'A searing account . . . raw and unflinching honesty' – *Daily Express* 'A genuine and courageous work of autobiography' – *Today*

A CHOICE OF PENGUINS AND PELICANS

Dinosaur and Co   Tom Lloyd

A lively and optimistic survey of a new breed of businessmen who are breaking away from huge companies to form dynamic enterprises in microelectronics, biotechnology and other developing areas.

The Money Machine: How the City Works   Philip Coggan

How are the big deals made? Which are the institutions that *really* matter? What causes the pound to rise or interest rates to fall? This book provides clear and concise answers to these and many other money-related questions.

Parkinson's Law   C. Northcote Parkinson

'Work expands so as to fill the time available for its completion': that law underlies this 'extraordinarily funny and witty book' (Stephen Potter in the *Sunday Times*) which also makes some painfully serious points for those in business or the Civil Service.

Debt and Danger   Harold Lever and Christopher Huhne

The international debt crisis was brought about by Western bankers in search of quick profit and is now one of our most pressing problems. This book looks at the background and shows what we must do to avoid disaster.

Lloyd's Bank Tax Guide 1988/9

Cut through the complexities! Work the system in *your* favour! Don't pay a penny more than you have to! Written for anyone who has to deal with personal tax, this up-to-date and concise new handbook includes all the important changes in this year's budget.

The Spirit of Enterprise   George Gilder

A lucidly written and excitingly argued defence of capitalism and the role of the entrepreneur within it.

Genetic Engineering for Almost Everybody   William Bains

Now that the 'genetic engineering revolution' has most certainly arrived, we all need to understand the ethical and practical implications of genetic engineering. Written in accessible language, they are set out in this major new book.

Brighter than a Thousand Suns   Robert Jungk

'By far the most interesting historical work on the atomic bomb I know of' – C. P. Snow

Turing's Man   J. David Bolter

We live today in a computer age, which has meant some startling changes in the ways we understand freedom, creativity and language. This major book looks at the implications.

Einstein's Universe   Nigel Calder

'A valuable contribution to the de-mystification of relativity' – *Nature*

The Creative Computer   Donald R. Michie and Rory Johnston

Computers *can* create the new knowledge we need to solve some of our most pressing human problems; this path-breaking book shows how.

Only One Earth   Barbara Ward and Rene Dubos

An extraordinary document which explains with eloquence and passion how we should go about 'the care and maintenance of a small planet'.

FOR THE BEST IN PAPERBACKS, LOOK FOR THE

A CHOICE OF PENGUINS AND PELICANS

Metamagical Themas   Douglas R. Hofstadter

A new mind-bending bestseller by the author of *Gödel, Escher, Bach*.

The Body   Anthony Smith

A completely updated edition of the well-known book by the author of *The Mind*. The clear and comprehensive text deals with everything from sex to the skeleton, sleep to the senses.

How to Lie with Statistics   Darrell Huff

A classic introduction to the ways statistics can be used to prove *anything*, the book is both informative and 'wildly funny' – *Evening News*

The Penguin Dictionary of Computers   Anthony Chandor and others

An invaluable glossary of over 300 words, from 'aberration' to 'zoom' by way of 'crippled lead-frog tests' and 'output bus drivers'.

The Cosmic Code   Heinz R. Pagels

Tracing the historical development of quantum physics, the author describes the baffling and seemingly lawless world of leptons, hadrons, gluons and quarks and provides a lucid and exciting guide for the layman to the world of infinitesimal particles.

The Blind Watchmaker   Richard Dawkins

'Richard Dawkins has updated evolution' – *The Times* 'An enchantingly witty and persuasive neo-Darwinist attack on the anti-evolutionists, pleasurably intelligible to the scientifically illiterate' – Hermione Lee in Books of the Year, *Observer*